T0227434

Quantum Physics
for Beginners

Quantum Physics

Physics

for Beginners

Zbigniew Ficek

PAN STANFORD PUBLISHING

Published by

Pan Stanford Publishing Pte. Ltd.
Penthouse Level, Suntec Tower 3
8 Temasek Boulevard
Singapore 038988

Email: editorial@panstanford.com
Web: www.panstanford.com

British Library Cataloguing-in-Publication Data
A catalogue record for this book is available from the British Library.

Quantum Physics for Beginners

Copyright © 2016 Pan Stanford Publishing Pte. Ltd.

ISBN 978-981-4669-38-2 (Hardcover)
ISBN 978-981-4669-39-9 (eBook)

Contents

Preface

Quantum mechanics is very puzzling.
A particle can be delocalized,
it can be simultaneously in several energy states
and it can even have several different identities at once.
                                        —Serge Haroche*

Quantum physics, also known as quantum mechanics or quantum wave mechanics—born in the late 1800s—is a study of the submicroscopic world of atoms, the particles that compose them and the particles that compose those particles. In 1800s, physicists believed that radiation is a wave phenomenon and matter is continuous; they believed in the existence of ether and had no ideas of what charge was. A series of the following experiments performed in late 1800s led to the formulation of quantum physics:

- Discovery of the electron
- Discovery of X-rays
- Observation of photoelectric effect
- Observation of discrete atomic spectra

Especially important among these was the interpretation of the spectrum of blackbody radiation that led to the breakdown of the equipartition theorem for electromagnetic radiation.

This textbook covers the background theory of various effects discussed from first principles, and as clearly as possible, to

---

*Haroche was granted the Nobel Prize in 2012 for ground-breaking experimental methods that enable measuring and manipulation of individual quantum systems.

introduce readers to the main ideas of quantum physics and to teach the basic mathematical methods and techniques used in the fields of advanced quantum physics, atomic physics, laser physics, nanotechnology, quantum chemistry, and theoretical mathematics. It also describes some of the key problems of quantum physics, concentrating on the background derivation, techniques, results, and interpretations. The book can be understood by a reader with little or even no previous knowledge of modern and quantum physics. It will help the readers learn how it comes about that microscopic objects (particles) behave in unusual ways called quantum effects, what the term "quantum" means, and where this idea came from. The book makes no attempt to a complete exploration of all predictions of quantum physics, but it is hoped that the predictions and problems explored in it will provide a useful starting point for those interested in learning more. It intends to explore problems that have been the most influential on the development of quantum physics and formulation of what we now call modern quantum physics. Many of the predictions of quantum physics appear to be contrary to our intuitive perceptions, and the goal to which this book aspires is a compact and logical exposition and interpretation of these fundamental and unusual predictions of quantum physics. Moreover, it contains numerous detailed derivations, proofs, worked examples, discussion problems, and a wide range of exercises from simple confidence-builders to fairly challenging problems, hard to find in textbooks on quantum physics. A number of problems, singled out as "tutorial problems", have been included in the chapters. It is important that the readers attempt these problems because they provide adequate understanding of the chapters. Another small set of the problems has also been introduced as "challenging problems", which are more advanced than the tutorial problems and are designed for readers with a passion for quantum physics.

The author would like to express his thanks to a number of students from the University of Queensland, King Saud University, and King Abdulaziz City for Science and Technology, who made use of a preliminary version of this book, for their valuable criticisms and suggestions for improvements.

Zbigniew Ficek
The National Centre for Applied Physics
King Abdulaziz City for Science and Technology
Riyadh, Saudi Arabia
Spring 2016

Note to the Reader

Quantum physics is the mathematical description of physical systems. A famous German physicist Arnold Sommerfeld* used to say, "If you want to study quantum physics, I give you three advices: (1) study mathematics, (2) study mathematics, (3) study more mathematics".

Calculus has been used extensively in the book, and therefore, the necessary prerequisites before attempting to study this book are that the reader should have some familiarity with wave mechanics, electromagnetism and optics, and a sufficient background in vector algebra, vector calculus, series and limits. In particular, complex numbers and functions of a complex variable, partial differentiation, multiple integrals, first- and second-order differential equations, Fourier series, matrix algebra, diagonalization of matrices, eigenvectors and eigenvalues, coordinate transformations, and special functions (Hermite and Legendre polynomials).

---

*Sommerfeld graduated in mathematics, worked for few years at some technical universities as an engineer, and next turned to physics. His greatest contribution to physics was the improvement of the Bohr model of a hydrogen atom. He included relativistic effects to the motion of an electron that allowed to explain the fine structure of the hydrogen spectrum. He also extended the model to elliptical orbits of the electron to describe the motion in terms of three quantum numbers what allowed to explain the Zeeman effect.

Suggested Readings

*Cohen-Tannoudji, C., Diu, B., and Laloe, F., *Quantum Mechanics* (Wiley, New York, 1977), Vols. I and II.

*Davydov, A. S., *Quantum Mechanics* (Pergamon Press, New York, 1965).

[1]Eisberg, R. and Resnick, R., *Quantum Physics of Atoms, Molecules, Solids, Nuclei, and Particles* (Wiley, New York, 1985).

[2]Krane, K., *Modern Physics* (Wiley, New York, 1996).

*Le Bellac, M., *Quantum Physics* (Cambridge University Press, Cambridge, 2006).

[3]Merzbacher, E., *Quantum Mechanics* (Wiley, New York, 1998).

*Messiah, A., *Quantum Mechanics*, (North-Holland, Amsterdam, 1962).

*Sakurai, J. J., *Modern Quantum Mechanics* (Addison-Wesley, 1994).

*Schiff, L., *Quantum Mechanics* (McGraw-Hill, New York, 1968).

[4]Serway, R. A., Moses, C. J., and Moyer, C. A, *Modern Physics* (Saunders, New York, 1989).

---

[1]Good introductory text on quantum physics with applications to atomic, molecular, solid state, and nuclear physics
[2]Good introductory text on quantum physics
[3]Excellent book on quantum physics
[4]Excellent introductory text on quantum physics
*Excellent books on quantum physics

Chapter 1

Radiation (Light) is a Wave

To anyone who is motivated by anything
beyond the most narrowly practical,
it is worthwhile to understand Maxwell's equations
simply for the good of his soul.

—J. R. Pierce

It is well known from classical optics that the most commonly observed phenomena with light (optical radiation) are polarization, interference, and diffraction. These phenomena are characteristic of waves, and some sort of wave theory is required for their explanation. Maxwell, in 1860, formulated the wave theory of light. He predicted the existence of electromagnetic waves that propagate with a speed, calculated from the laws of electricity and magnetism, similar to the measured speed of light. The prediction of electromagnetic waves and the subsequent successful use of Maxwell's theory in explaining the interference and diffraction phenomena made the theory to be recognized as the fundamental wave theory of radiation.

We begin our journey through the fundamentals of quantum physics with an elementary, but quantitative, classical theory of the radiative field. We first briefly outline the electromagnetic theory of radiation, and describe how the electromagnetic radiation may be understood as a wave that can be represented by a set of harmonic

Quantum Physics for Beginners
Zbigniew Ficek
Copyright © 2016 Pan Stanford Publishing Pte. Ltd.
ISBN 978-981-4669-38-2 (Hardcover), 978-981-4669-39-9 (eBook)
www.panstanford.com

oscillators. This is followed by a description of the Hamiltonian of the electromagnetic field, which determines the energy of the electromagnetic wave. In particular, we will be interested in how the energy of the electromagnetic wave depends on the parameters characteristic of the wave: amplitude and frequency. In our analysis, we assume that the readers are familiar with vector algebra and with the basic concepts of electricity and magnetism.

1.1 Wave Equation

We start by considering the time-varying electric ($\vec{E}$) and magnetic ($\vec{B}$) fields that propagate in an empty space, i.e., in the space that does not contain material bodies, but there can be free charges and currents present. The fields propagating in the space satisfy Maxwell's equations, the fundamental (experimental) laws relating to the electric and magnetic field vectors assembled together into a single set of differential equations[a]

$$\nabla \cdot \vec{E} = \rho/\varepsilon_0 \,, \tag{1.1}$$

$$\nabla \cdot \vec{B} = 0 \,, \tag{1.2}$$

$$\nabla \times \vec{E} = -\frac{\partial}{\partial t}\vec{B} \,, \tag{1.3}$$

$$\nabla \times \vec{B} = \mu_0 \vec{J} + \frac{1}{c^2}\frac{\partial}{\partial t}\vec{E} \,, \tag{1.4}$$

where ρ is the density of free charges and $\vec{J}$ is the density of currents at a point where the electric and magnetic fields are evaluated. The parameters ε_0 and μ_0 are constants that determine the property of the vacuum and are called the electric permittivity and magnetic permeability, respectively. The parameter $c = 1/\sqrt{\varepsilon_0\mu_0}$, and its numerical value is equal to the speed of light in vacuum, $c = 3 \times 10^8$ (ms^{-1}).

The symbol ∇ is called "nabla" or "del." It is a vector differential operator and, in the Cartesian coordinates, has the form

$$\nabla = \vec{i}\frac{\partial}{\partial x} + \vec{j}\frac{\partial}{\partial y} + \vec{k}\frac{\partial}{\partial z} \,, \tag{1.5}$$

---

[a]Readers familiar with the laws of electricity and magnetism should recognize which of the fundamental laws of electromagnetism are assembled together to form Maxwell's equations.

where $\vec{i}$, $\vec{j}$, and $\vec{k}$ are the unit vectors in the orthogonal x, y, and z directions, respectively. We have much more to say about this operator, but for the present it is sufficient to note that nabla is incomplete as it stands; it needs something to operate on. In other words, nabla may be viewed formally as an operator converting a scalar function into a vector. The dot ($\cdot$) and the cross ($\times$) symbols appearing in Maxwell's equations are the standard scalar and vector products between two vectors, respectively.

The fact that the numerical value of c is 3×10^8 [ms^{-1}], which is the velocity of light in vacuum, led Maxwell to propose an electromagnetic theory of light, one of the brilliant contributions to physics of the 19th century.

Let us analyze briefly Maxwell's idea leading to the electromagnetic theory of light. Suppose, we have electric and magnetic fields in the absence of free currents and charges, $\vec{J} = 0$, $\rho = 0$. In this case, Maxwell's equations (1.1)–(1.4) describe a free electromagnetic field, i.e., an electromagnetic field in vacuum. Of course, there must be a source of the fields somewhere, but the region in which we consider the fields is free of currents and charges.

With $\vec{J} = 0$ and $\rho = 0$, we may reduce Maxwell's equations into two differential equations for $\vec{E}$ or $\vec{B}$ alone. To show this, we apply $\nabla\times$ to both sides of Eq. (1.3), and then using Eq. (1.4), we find

$$\nabla \times (\nabla \times \vec{E}) = -\frac{\partial}{\partial t}(\nabla \times \vec{B}) = -\frac{1}{c^2}\frac{\partial^2}{\partial t^2}\vec{E} . \qquad (1.6)$$

Since

$$\nabla \times (\nabla \times \vec{E}) = -\nabla^2\vec{E} + \nabla(\nabla \cdot \vec{E}), \qquad (1.7)$$

and $\nabla \cdot \vec{E} = 0$ in the vacuum, we obtain

$$\nabla^2\vec{E} - \frac{1}{c^2}\frac{\partial^2}{\partial t^2}\vec{E} = 0, \qquad (1.8)$$

where the operator

$$\nabla^2 = \nabla \cdot \nabla = \frac{\partial^2}{\partial x^2} + \frac{\partial^2}{\partial y^2} + \frac{\partial^2}{\partial z^2} \qquad (1.9)$$

is called Laplacian and is a scalar.

Readers familiar with harmonic motion will recognize that Eq. (1.8) is a form of the well-known differential equation in physics, called the *Helmholtz wave equation* for a harmonic motion. It is in the standard form of a three-dimensional vector wave equation.

Similarly, we can derive a wave equation for the magnetic field, and the readers may readily verify that field $\vec{B}$ satisfies the same equation as $\vec{E}$:

$$\nabla^2 \vec{B} - \frac{1}{c^2} \frac{\partial^2}{\partial t^2} \vec{B} = 0 . \qquad (1.10)$$

The wave equations (1.8) and (1.10) can be solved with a minor difficulty, and the general solution is given in a form of a superposition of plane harmonic waves, specified by the symbol k as

$$\vec{U} = \sum_k \vec{U}_k\, e^{-i\left(\omega_k t - \vec{k}\cdot\vec{r}\right)} , \qquad (1.11)$$

where $\vec{U} \equiv (\vec{E}, \vec{B})$, the parameter ω_k is the frequency of the kth wave, and $\vec{U}_k$ is the amplitude of the $\vec{E}$ or $\vec{B}$ wave propagating in the $\vec{k}$ direction. The solution (Eq. 1.11) represents a set of harmonic waves of frequencies ω_k and propagating with a velocity c in the $\vec{k}$ direction relative to the position $\vec{r}$ of an observation point.

The vector $\vec{k}$ is called the *wave vector* that, as we have already mentioned, determines the direction of propagation of the kth wave. From the requirement that Eq. (1.11) is a solution to the wave equation (Eq. 1.8), we find that the magnitude of $\vec{k}$ as: $|\vec{k}| = \omega_k/c = 2\pi/\lambda_k$, where λ_k is the wavelength of the kth wave.[a]

Summarizing briefly: Maxwell's theory of electromagnetism shows that the electric and magnetic fields propagate in vacuum as plane (electromagnetic) waves.

The electromagnetic waves have specific properties, which are determined from Maxwell's equations. It is easy to show that the divergence Maxwell's equations (1.1) and (1.2) demand that for all directions of propagation $\vec{k}$:

$$\vec{k} \cdot \vec{E}_k = 0 \qquad \text{and} \qquad \vec{k} \cdot \vec{B}_k = 0 . \qquad (1.12)$$

This means that the electric and magnetic fields are both perpendicular to the direction of propagation $\vec{k}$, i.e., both $\vec{E}$ and $\vec{B}$ have no

---

[a]The relationship between wavelength, velocity, and frequency obtained here is undoubtedly familiar to the readers who have undertaken courses in physics.

components in the direction of propagation. Such a wave is called a *transverse wave*.

Maxwell's equations (1.3) and (1.4) provide a further restriction on the fields that

$$\vec{B}_k = \frac{1}{c}\vec{\kappa} \times \vec{E}_k,$$ (1.13)

where $\vec{\kappa} = \vec{k}/|\vec{k}|$ is the unit vector in the $\vec{k}$ direction. This equation shows that the electric ($\vec{E}$) and magnetic ($\vec{B}$) fields of an electromagnetic wave propagating in vacuum are mutually orthogonal ($\vec{E} \perp \vec{B}$).

In summary: From the foregoing, we see that only transverse waves are predicted in vacuum, both electric and magnetic fields being perpendicular to the direction of propagation and to each other.

1.2 Energy of the Electromagnetic Wave

Electromagnetic waves contain and transport energy. The energy of an electromagnetic wave is carried in its electric and magnetic fields, a result we could well anticipate in view of our knowledge of the properties of electric and magnetic fields.

In order to determine the energy carried by an electromagnetic wave and how the energy depends on the parameters characteristic of the wave (amplitude and frequency), we consider a simple example of a plane electromagnetic wave propagating in one dimension, say the positive z-axis: $\vec{k} \cdot \vec{r} = kz$. Suppose that the wave is linearly polarized in the x-direction. In this case, the wave may be determined by the electric field

$$\vec{E}(z, t) = \vec{i}E_x(z, t) = \vec{i}q(t)\sin(kz),$$ (1.14)

where $q(t)$ represents the time-dependent amplitude of the electric field.

The energy of the electromagnetic wave is given by the Hamiltonian

$$H = \frac{1}{2}\int_0^L dz \left\{ \varepsilon_0 |\vec{E}|^2 + \frac{1}{\mu_0}|\vec{B}|^2 \right\},$$ (1.15)

where $\varepsilon_0|\vec{E}^2|/2$ is the density of energy stored in the electric field, and $|\vec{B}|^2/(2\mu_0)$ is the density of energy stored in the magnetic field.

The integration is to be carried out over the whole length L where the wave exists.

One can see from Eq. (1.15) that to determine the energy of the electromagnetic wave, we need the magnetic field $\vec{B}$. Since we know the electric field, we can find the magnetic field from the Maxwell equation (1.4). For the electromagnetic wave, the magnetic vector $\vec{B}$ is perpendicular to $\vec{E}$ and oriented along the y-axis. Hence, the magnitude of the magnetic field can be found from the following equation:

$$\nabla \times \vec{B} = \vec{i}\,\frac{1}{c^2}\dot{q}\,(t)\sin(kz)\,. \tag{1.16}$$

Since $B_x = B_z = 0$ and $B_y \neq 0$, and obtain

$$\nabla \times \vec{B} = -\vec{i}\,\frac{\partial B_y}{\partial z} + \vec{k}\,\frac{\partial B_y}{\partial x} = \vec{i}\,\frac{1}{c^2}\dot{q}\,(t)\sin(kz)\,. \tag{1.17}$$

The coefficients on both sides of Eq. (1.17) at the same unit vectors should be equal. Hence, we find that

$$\frac{\partial B_y}{\partial x} = 0 \quad \text{and} \quad \frac{\partial B_y}{\partial z} = -\frac{1}{c^2}\dot{q}\,(t)\sin(kz)\,. \tag{1.18}$$

Then, integrating $\partial B_y/\partial z$ over z, we find

$$B_y\,(z, t) = -\frac{1}{c^2}\dot{q}\,(t)\int dz\sin(kz) = \frac{1}{kc^2}\dot{q}\,(t)\cos(kz)\,. \tag{1.19}$$

Thus, the energy of the electromagnetic field, given by the Hamiltonian (1.15), is of the form

$$H = \frac{1}{2}\int_0^L dz\left\{\varepsilon_0 q^2\,(t)\sin^2(kz) + \frac{1}{k^2c^4\mu_0}\,(\dot{q}\,(t))^2\cos^2(kz)\right\}. \tag{1.20}$$

Since

$$\int_0^L dz\sin^2(kz) = \int_0^L dz\cos^2(kz) = \frac{1}{2}L\,, \tag{1.21}$$

and $\mu_0 = 1/c^2\varepsilon_0$, the Hamiltonian H reduces to

$$H = \frac{1}{4}\varepsilon_0 Lq^2\,(t) + \frac{1}{4}\frac{\varepsilon_0}{\omega^2}L\,(\dot{q}\,(t))^2\,. \tag{1.22}$$

An inspection of this equation leads us to conclude that the energy of the electromagnetic wave is proportional to the square of its amplitude, $q\,(t)$.

Note that the Hamiltonian (1.22) is in the form of the familiar Hamiltonian of a harmonic oscillator. It is easy to prove. We know from classical mechanics that the energy of a harmonic oscillator is given by the sum of the kinetic and potential energies as

$$H_{\text{osc}} = E_K + E_p = \frac{1}{2m}\left(p^2 + m^2\omega^2 x^2\right), \qquad (1.23)$$

where $p = m\dot{x}$ is the momentum of the oscillating mass m. Comparing Eq. (1.23) with Eq. (1.22), we see that the variables $q(t)$ and $\dot{q}(t)$ can be related to the position and momentum of the harmonic oscillator.

Summary

We have learned that

(1) The electromagnetic field propagates in vacuum as transverse plane waves, which can be represented by a set of harmonic oscillators. Thus, according to Maxwell's electromagnetic theory, radiation (light) is a wave.
(2) The energy (intensity) of the electromagnetic field is proportional to the square of the amplitude of the oscillation.

Worked Example

Show that the single-mode electric field

$$\vec{E} = \vec{E}_0 \sin\left(k_x x\right) \sin\left(k_y y\right) \sin\left(k_z z\right) \sin\left(\omega t + \phi\right) \qquad (1.24)$$

is a solution to the wave equation (1.8) if $k = \omega/c$, where $k = \left(k_x^2 + k_y^2 + k_z^2\right)^{\frac{1}{2}}$ is the magnitude of the wave vector, and ϕ is an arbitrary initial phase of the propagating field.

Solution

To show that the electric field (1.24) satisfies the wave equation, we have to calculate the time and coordinate second derivatives of $\vec{E}$.

Thus, using Eq. (1.24), we find

$$\frac{\partial^2 \vec{E}}{\partial t^2} = -\omega^2 \vec{E} \,, \tag{1.25}$$

and

$$\frac{\partial^2 \vec{E}}{\partial x^2} = -k_x^2 \vec{E} \,, \qquad \frac{\partial^2 \vec{E}}{\partial y^2} = -k_y^2 \vec{E} \,, \qquad \frac{\partial^2 \vec{E}}{\partial z^2} = -k_z^2 \vec{E} \,. \tag{1.26}$$

Hence, substituting Eqs. (1.25) and (1.26) into the wave equation

$$\left(\frac{\partial^2}{\partial x^2} + \frac{\partial^2}{\partial y^2} + \frac{\partial^2}{\partial z^2} \right) \vec{E} - \frac{1}{c^2} \frac{\partial^2 \vec{E}}{\partial t^2} = 0 \,, \tag{1.27}$$

we obtain

$$- \left(k_x^2 + k_y^2 + k_z^2 \right) \vec{E} + \frac{\omega^2}{c^2} \vec{E} = 0 \,, \tag{1.28}$$

or

$$\left[\left(k_x^2 + k_y^2 + k_z^2 \right) - \frac{\omega^2}{c^2} \right] \vec{E} = 0 \,. \tag{1.29}$$

Since $\vec{E} \neq 0$ and $k_x^2 + k_y^2 + k_z^2 = k^2$, we find that the left-hand side of Eq. (1.29) is equal to zero when

$$k^2 = \frac{\omega^2}{c^2} \,, \qquad \text{i.e. when} \qquad k = \frac{\omega}{c} \,. \tag{1.30}$$

Hence, the single-mode electric field (Eq. 1.24) is a solution to the wave equation if $k = \omega/c$.

Revision Questions

Question 1. What is a wave equation and how it is derived from Maxwell's equations?

Question 2. What is the solution to a wave equation?

Question 3. What are the properties of an electromagnetic wave propagating in vacuum?

Question 4. Energy of an electromagnetic wave is proportional to the amplitude of the wave: true or false?

Tutorial Problems

Problem 1.1 Verify that the general solution (1.11) satisfies the wave equation (1.8).

Problem 1.2 Using Eq. (1.13), show that

$$\vec{E}_k = -c\vec{\kappa} \times \vec{B}_k ,$$

which is the same relation one could obtain from the Maxwell equation (1.4).
(*Hint:* Use the vector identity $\vec{A} \times (\vec{B} \times \vec{C}) = \vec{B}(\vec{A} \cdot \vec{C}) - \vec{C}(\vec{A} \cdot \vec{B})$.)

Problem 1.3 Using the divergence Maxwell equations, show that the electromagnetic waves in vacuum are transverse waves.

Problem 1.4 Calculate the energy of an electromagnetic wave propagating in one dimension.

Chapter 2

Difficulties of the Wave Theory of Radiation

We have already seen how the fundamental laws of electromagnetism led to the prediction that light is an electromagnetic wave. But what properties of light are commonly observed in experiments? Since light has been recognized as a wave phenomenon, one can say that we shall observe properties characteristic of waves. Classical optics provides us with many examples of phenomena reflecting the wave character of light. Typical are polarization, interference, and diffraction phenomena. However, a series of experiments performed in the late 19th century showed that the wave model predicted from Maxwell's equations is not the correct description of the properties of light. In this chapter, we will discuss some of the experiments that provided evidence that light, which we have recognized as a wave phenomenon, has properties that we normally associate with particles. In particular, these experiments indicated that in some phenomena, the energy of light does not show up as being proportional to the amplitude of the oscillation; it is rather proportional to the frequency of the oscillation. The experimental results will force us to conclude that the wave theory of light is simply inappropriate to explain these phenomena. We will discover

Quantum Physics for Beginners
Zbigniew Ficek
Copyright © 2016 Pan Stanford Publishing Pte. Ltd.
ISBN 978-981-4669-38-2 (Hardcover), 978-981-4669-39-9 (eBook)
www.panstanford.com

later that to explain these phenomena and to understand the physical processes involved, some kind of a corpuscular theory of light is required.

2.1 Discovery of Electron

Let us first discuss briefly how the idea of the existence of discrete (corpuscular) structures has been innovated in physics. The first idea of the existence of discrete structures referred to physics of materials and their electric properties.[a] Joseph Thomson, in his famous q/m experiment performed in 1896, measured the ratio of the charge q to the mass m of the cathode beam particles.[b] He found that the ratio q/m

- Did not depend on the cathode material.
- Did not depend on the residual gas in the tube.
- Did not depend on anything else about the experiment.

This independence, discovered by Thomson, showed that the particles in the cathode beam are a common element of all materials.[c]

In 1910–13, Robert Millikan performed a series of oil-drop experiments, in which he measured electric charge of individual oil drops.[d] In these experiments, he made an important observation[e] that every drop had a charge q equal to some small *integer* multiple of a basic (elementary) charge e ($q = ne$), where n is an integer ($n = 1, 2, \ldots$), and $e = 1.602177 \times 10^{-19}$ [C].

---

[a] In fact, the idea of the corpuscular nature of the physical world, particularly the corpuscular nature of light, was first introduced by Newton. It was abounded when the wave properties of light were observed experimentally.

[b] For a complete discussion of Thomson's experiment see, for example, R.A. Serway, C.J. Moses, and C.A. Moyer, *Modern Physics* (Saunders, New York, 1989), p. 80.

[c] Thomson was granted the Nobel prize in 1906 for his theoretical and experimental investigations on the conduction of electricity by gases.

[d] For a complete discussion of Millikan's experiment see, for example, R.A. Serway, C.J. Moses, and C.A. Moyer, *Modern Physics* (Saunders, New York, 1989), p. 83.

[e] Millikan was granted the Nobel Prize in 1923 for his work on the elementary charge of electricity.

Thus, we may summarize that Thomson's and Millikan's experiments clearly indicate that matter is not continuous but composed of discrete particles. The most important discovery was the existence of particles that are a common element of all materials! These particles are *electrons*, and Millikan's experiments showed that each electron carries the same amount of electric charge. Physicists believe, but not all are convinced, that electrons do not have internal structure and are the smallest particles (elementary particles) that exist in the universe. Why there exist the smallest particle remains, however, one of the unsolved mysteries of physics.

2.1.1 Discovery of X-Rays

In 1895, Wilhelm Röntgen was interested in the study of the passage of a cathode beam through an aluminum-foil window. He made a classic observation that a highly penetrating radiation of unknown nature was produced when the cathode beam impinged on matter. He observed that the radiation traveled in straight lines even through electric and magnetic fields. He also found that the radiation passed readily through opaque materials and exposed photographic plates for no apparent reason. He called this unknown radiation X-rays.[a]

In 1906, Charles Barkla[b] observed a partial polarization of X-rays, which indicated that they were transverse waves.

If X-rays are transverse waves and are invisible, then an obvious question arises: What are the wavelengths of X-rays?

To answer this question, let us think how we can measure the wavelength of X-rays. One possibility, in principle at least, would be to perform Young's double-slit experiment. However, any attempt to measure the wavelength using Young's double-slit experiment was unsuccessful with no interference pattern observed.

In 1912, Max Laue got an idea that resolved this problem. He explained that no interference pattern was observed simply because the wavelengths of X-rays were too small. To explore his argument,

---

[a] Röntgen was granted the Nobel Prize in 1901 for his discovery of X-rays.
[b] Barkla was granted the Nobel Prize in 1917 for his discovery of the characteristic X-rays of elements.

consider the condition for observation of an interference pattern

$$2d \sin \theta_n = n\lambda, \qquad n = 0, 1, 2, \ldots, \tag{2.1}$$

from which we have

$$\sin \theta_n = \frac{n\lambda}{2d}, \tag{2.2}$$

where λ is the wavelength of the X-rays, d is the separation between the slits, and the integer n numbers the successive maxima or minima in the interference pattern. For $\lambda \ll d$, we have $\sin \theta_n \approx 0$ even for large n. Hence, to make $\sin \theta_n \approx 1$ to see the interference fringes separated from each other, the separation d between the slits should be very small.

Laue proposed to use a crystal for the interference experiment.[a] This Nobel Prize winning idea was based on the following observation: In crystals, the average separation between the atoms, acting as slits, is about $d \approx 0.1$ nm. From the experiment, he found that the wavelength of the measured X-rays was $\lambda \approx 0.6$ nm. Typical X-ray wavelengths are in the range 0.001–1 nm. These are very short wavelengths well outside the ultraviolet wavelengths. For a comparison, visible light wavelengths are between $\lambda \approx 410$ nm (violet) and $\lambda \approx 656$ nm (red).

We know how to generate visible light, but how this invisible radiation of such small wavelengths is generated?

This can be explained as follows: X-rays are generated when high-speed electrons crash into the anode and rapidly deaccelerate. It is well known from the theory of classical electrodynamics that electrons, when deaccelerated, emit radiation. In other words, their kinetic energy is converted into radiation energy (braking, i.e., deaccelerating radiation, often referred to by the German phrase bremsstrahlung).

The total instantaneous power P radiated by the deaccelerated electron is given by the Larmor formula:

$$P = \frac{2}{3} \frac{e^2}{4\pi \varepsilon_0 c^3} |a|^2, \tag{2.3}$$

where $|a|$ is the magnitude of the deacceleration of the electron.

---

[a] Laue was granted the Nobel Prize in 1914 for his discovery of the diffraction of X-rays by crystals.

Hence, due to the continuous deacceleration of the electrons, the spectrum $I(\lambda)$ of X-rays also should be continuous. However, the experimentally observed spectrum of the X-rays is composed of two pronounced peaks at certain wavelengths superimposed on a continuous background (see Fig. 2.1). It is observed that the position of the two peaks depends only on the material of the anode (characteristic radiation). The origin of the two lines is unknown!

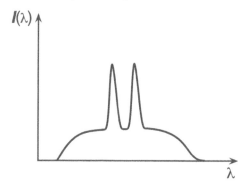

Figure 2.1 An example of experimentally observed spectrum $I(\lambda)$ of X-ray radiation.

Yet another unexpected property of the X-rays spectrum is as follows. For a given potential V, experiments showed that X-rays of different wavelengths were produced, giving the continuous background seen in Fig. 2.1, but none of the wavelengths was shorter than a certain wavelength λ_{min}. The minimum wavelength λ_{min} was observed to depend only on the potential in the tube $(\lambda_{min} \sim V)$ and was the same for all target (anode) material. The reason was unknown!

2.2 Photoelectric Effect

In 1887, Heinrich Hertz discovered the photoelectric effect: emission of electrons from a surface (cathode) when light strikes on it. If a positively charged electrode is placed near the photoemissive cathode to attract photoelectrons, an electric current can be made to flow in response to the incident light.

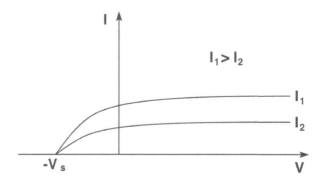

Figure 2.2 Photoelectric effect for two different intensities I_1 and I_2 ($I_1 > I_2$) of the incident light.

The following properties of the photoelectric effect were observed:

1. When a monochromatic light falls on the cathode, no electrons are emitted, regardless of the intensity of the light, unless the *frequency* (not the intensity!) of the incident light is high enough to exceed some minimum value, called the threshold frequency. The threshold frequency was found to be dependent on the material of the cathode.
2. Once the frequency of the incident light is greater than the threshold value, some electrons are emitted from the cathode with a nonzero speed. The reversed potential is required to stop the electrons (stopping potential: $eV_s = \frac{1}{2}mv^2$).
3. When the intensity of light is increased, while its frequency is kept the same, more electrons are emitted, but the stopping potential V_s is the same (see Fig. 2.2).
   Conclusion: Velocity of the electrons, which is proportional to the energy, is unaffected by changes in the intensity of the incident light.
4. When the frequency of light is increased, $v_2 > v_1$, the stopping potential increased, $V_{s2} > V_{s1}$ (see Fig. 2.3).

In summary: We see that the results of the experiments on photoelectric effect are in contradiction to classical wave theory.

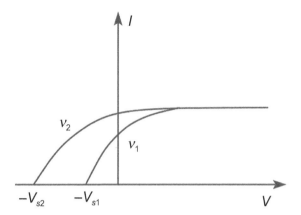

Figure 2.3 Photoelectric effect for two different frequencies but the same intensity of the incident light, with $\nu_2 > \nu_1$.

These puzzling features of the photoelectric effect suggest that the energy of light is *not* proportional to its intensity but is proportional to the frequency:

$$E \sim \nu \quad \text{or} \quad E \sim \frac{1}{\lambda}, \quad \left(\nu = \frac{c}{\lambda}\right). \quad (2.4)$$

It is impossible to explain the aspects of the photoelectric effect by means of the wave theory of light that gives no reason why the energy of a light beam should be proportional to frequency. The wave theory of light leads one to anticipate that a long-wavelength light incident on a surface could cause enough energy to be absorbed for an electron to be released. Moreover, when electrons are emitted, an increase in the incident light intensity should cause an emitted electron to have more kinetic energy rather than more electrons of the same average energy to be emitted.

2.3 Compton Scattering

Compton scattering experiments, named after Arthur Compton who, in 1924, performed a series of experiments on scattering of

X-rays on free electrons,[a] provided additional direct confirmation that energy of light is proportional to its frequency rather than to the amplitude. In the Compton experiment, the light of a wavelength λ was scattered on free electrons, as shown in Fig. 2.4.

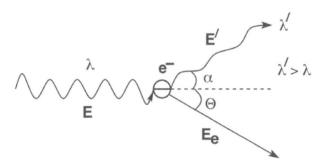

Figure 2.4   Schematic diagram of the Compton scattering. An incident light of wavelength λ is scattered on free electrons and the scattered light of wavelength λ' is detected in the α direction.

It was observed that during the scattering process, the intensity of the incident light did not change, but the *wavelength* changed such that the wavelength of the scattered light was larger than the incident light, i.e., $\lambda' > \lambda$.

Conclusion: From energy conservation, we have

$$E = E' + E_e, \tag{2.5}$$

where E_e is the energy of the scattered electrons.
Since $E_e > 0$, $E' < E$, indicating that the energy of the incident light is proportional to the frequency, or equivalently, to the inverse of the wavelength

$$E \sim \nu \quad \text{or} \quad E \sim \frac{1}{\lambda}. \tag{2.6}$$

This is another notable departure from the predictions of the wave theory of radiation.

---

[a]Compton was granted the Nobel Prize in 1927 for his discovery of the effect named after him.

2.4 Discrete Atomic Spectra

In a different area of physics, radiation spectroscopy experiments show that light emitted by a hot solid or liquid exhibits a continuous spectrum, i.e., light of all wavelengths is emitted. In 1814, Joseph von Fraunhoffer observed that the continuous spectrum of light coming from the sun also contained some discrete (black) lines that were later recognized as absorption lines corresponding to the elements in the outer atmosphere of the sun. Moreover, experiments showed that the spectrum of light emitted by a hot gas contained only a few isolated sharp lines (see Fig. 2.5) of the following properties:

Figure 2.5 An example of discrete radiation spectrum emitted from single atoms.

- Each spectral line corresponds to a different frequency.
- The lines group into separate sequences, called spectral series.
- Different gases produce different sets of lines.
- When we increase the temperature of a gas, more lines at larger frequencies are emitted.

Once again, we are faced with the difficulty of explaining experimental observations using the wave theory of light. Evidently, these results contradict the prediction of the wave theory of light since the experimentally observed spectra show that energy is proportional to frequency, $E \sim \nu$, not to the amplitude of the emitted light. Moreover, this shows that the structure of atoms is not continuous.

Then additional questions arise: What is an atom composed of? How does the discrete spectrum relate to the internal structure of the atom?

These questions were left without answers at that time.

Summary

A series of experiments on properties of X-rays, properties of photoelectric effect, Compton scattering and atomic spectra suggest that energy of the radiation field (light) is not proportional to its amplitude, as one could expect from the wave theory of light, but to the frequency of the radiation field, $E \sim \nu$.

In the next chapter, we consider one more example of unusual experimental results, the blackbody radiation, which solved the problem.

Chapter 3

Blackbody Radiation

The radiation emitted by a body that is heated is called *thermal radiation*. All bodies emit and absorb such radiation. We have just seen that hot gases or individual atoms emit radiation with characteristic discrete lines. In contrast, hot matter in a condensate state (solid or liquid) emits radiation with a continuous distribution of wavelengths rather than a line spectrum.

In this chapter, we will consider spectral distribution of the continuous radiation emitted by a blackbody, i.e., the dependence of the intensity of the blackbody radiation on the wavelength of the radiation.[a] First, we will define what we mean by a blackbody. *Blackbody* is an object that absorbs completely all radiation falling on it, independent of its frequency, wavelength, and intensity. In other words, no reflections occur from a blackbody.

Examples: Consider a box with perfectly reflecting sides and with a small hole, as illustrated in Fig. 3.1. The small hole, not the box itself, can be treated as a blackbody. Radiation of an arbitrary wavelength

---

[a]In spectroscopy, the spectral distribution (or spectrum) of a radiation is usually determined by the dependence of the radiation intensity on frequency rather than wavelength. Since the blackbody spectrum was first analyzed in terms of wavelength, we will follow this historical development.

Quantum Physics for Beginners
Zbigniew Ficek
Copyright © 2016 Pan Stanford Publishing Pte. Ltd.
ISBN 978-981-4669-38-2 (Hardcover), 978-981-4669-39-9 (eBook)
www.panstanford.com

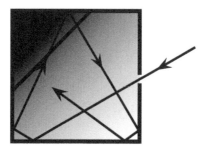

Figure 3.1 A box with a small hole that models a blackbody.

that strikes on the hole gets lost inside the box. Thus, no reflection occurs from the hole, which is the property of a blackbody.

Formulation of the correct theoretical approach to calculate the spectrum of the blackbody radiation was a major challenge in theoretical physics during the late 19th century. In 1900, Strutt (Lord) Rayleigh calculated the energy density distribution $I(\lambda)$ of the radiation emitted by the blackbody box at absolute temperature T.

According to Rayleigh, the energy density distribution (spectrum) of the blackbody radiation is of the form

$$I(\lambda) = N_\lambda \langle E \rangle , \tag{3.1}$$

where N_λ is the number of radiating oscillators (modes), per unit volume and unit wavelength, inside the box[a]

$$N_\lambda = \frac{8\pi}{\lambda^4} , \tag{3.2}$$

and $\langle E \rangle$ is the average energy of each mode. Note that $\langle E \rangle$ in the Rayleigh and Jeans formula is the same for each mode. This is consistent with the wave theory of radiation that the energy of a radiation wave is proportional to its amplitude, not to the wavelength λ.

---

[a]The formula for the density of modes derived by Rayleigh missed the factor 8, which was corrected by James Jeans in 1905. Therefore, the energy density distribution (3.1) is called the Rayleigh–Jeans formula.

Before discussing the Rayleigh–Jeans formula in details, we will first prove Eq. (3.2) for the number of modes inside the box, and next we will find the average energy of each mode in terms of the absolute temperature T.

3.1 Number of Radiation Modes inside a Box

Consider an electromagnetic wave confined in the volume V. We take a plane wave propagating in the $\vec{r}$ direction, which in terms of x, y, z components can be written as

$$\vec{E} = \vec{E}_0 \sin\left(k_x x\right) \sin\left(k_y y\right) \sin\left(k_z z\right) \sin\left(\omega t + \phi\right) . \qquad (3.3)$$

The wave propagating in the box interferes with the waves reflected from the walls. The interference will destroy the wave unless it forms a standing wave inside the box. The wave forms a standing wave when the amplitude of the wave vanishes at the walls. This happens when

$$\sin\left(k_x x\right) = 0 , \quad \sin\left(k_y y\right) = 0 , \quad \sin\left(k_z z\right) = 0 , \qquad (3.4)$$

i.e., when

$$k_x = \frac{n\pi}{x} , \quad k_y = \frac{m\pi}{y} , \quad k_z = \frac{l\pi}{z} , \qquad (3.5)$$

where n, m, l are integer numbers $(n, m, l = 1, 2, 3, \ldots)$.

The conditions (3.5) are called the **boundary condition**, i.e., condition imposed on the wave at the walls to form standing waves inside the box. The standing wave condition is common to all confined waves. In vibrating violin strings or organ pipes, for example, it also happens that only those frequencies that satisfy the aforementioned boundary condition are permitted. Thus, inside the box, the electromagnetic field is a combination of standing waves.

Since $k = 2\pi/\lambda$, we have the following condition for a standing wave inside the box in terms of the wavelength

$$x = \frac{n\lambda}{2} , \quad y = \frac{m\lambda}{2} , \quad z = \frac{l\lambda}{2} , \qquad (3.6)$$

We see that each set of the numbers (n, m, l) corresponds to a particular wave, which we call *mode*.

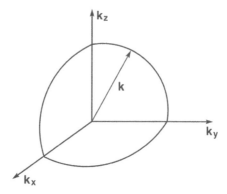

Figure 3.2 Illustration of the positive octant of k-space. The dots represent modes of different (n, m, l). In the limit of a large number of modes, the roughness of the edges is insignificant. Therefore, we can assume a continuous distribution of the modes and approximate it by a sphere.

To find the number of modes inside the box in terms of frequency or wavelength of the radiation, it is convenient to work in the propagation k-space spanned on the components k_x, k_y, and k_z of the $\vec{k}$ vector, as shown in Fig. 3.2. Each set of the components k_x, k_y, k_z is represented by the numbers n, m, l.

We will calculate the number of the modes as the ratio of the volume occupied by all the modes to the volume occupied by a single mode.

In the k-space, a single mode, say $(n, m, l) = (1, 1, 1)$, occupies a volume

$$V_k = k_x k_y k_z = \frac{\pi^3}{xyz} = \frac{\pi^3}{V},$$ (3.7)

where $V = xyz$ is the volume of the box.

Since k_x, k_y, k_z are positive numbers, the modes propagate only in the positive octant of the k-space. The number of modes inside the octant, shown in Fig. 3.2, is given by

$$N(k) = \frac{1}{8} \frac{\frac{4}{3}\pi k^3}{V_k},$$ (3.8)

i.e., it is equal to the volume of the octant occupied by all the modes divided by the volume occupied by a single mode. Since $k = 2\pi v/c$, we get

$$N(k) = \frac{8\pi \nu^3}{3c^3} V \,, \tag{3.9}$$

where we have increased $N(k)$ by a factor of 2. This arises from the fact that the light is assumed to be generated by a thermal field and hence is completely unpolarized. Thus, it may be regarded as a mixture of waves of two mutually orthogonal polarizations.

In the limit of a large number of modes, we can assume a continuous distribution of the modes, and then the number of modes per unit volume and per unit frequency needed for the energy density distribution is

$$N_\nu = \frac{1}{V} \frac{dN(k)}{d\nu} = \frac{8\pi \nu^2}{c^3} \,. \tag{3.10}$$

In terms of wavelengths λ, the number of modes per unit volume and per unit of wavelength is given by the following conversion formula:

$$N_\lambda = \frac{1}{V} \frac{dN(k)}{d\lambda} = \frac{1}{V} \frac{dN(k)}{d\nu} \left| \frac{d\nu}{d\lambda} \right| = N_\nu \left| \frac{d\nu}{d\lambda} \right| \,. \tag{3.11}$$

Since $\nu = c/\lambda$ and $d\nu = (-c/\lambda^2)d\lambda$, we obtain

$$N_\lambda = \frac{8\pi}{\lambda^4} \,. \tag{3.12}$$

This is the final formula for the number of modes (allowed standing waves) inside the box. Note that the number of modes is determined in terms of wavelength, and Eq. (3.11) must be kept in mind when translating results for the mode density from a wavelength to a frequency scale.

3.2 Average Energy of a Radiation Mode: Equipartition Theorem

We now return to the Rayleigh and Jeans calculations of the energy density distribution $I(\lambda)$ of the blackbody radiation:

$$I(\lambda) = N_\lambda \langle E \rangle \,. \tag{3.13}$$

The average energy of a radiation mode that appears in Eq. (3.13) may be found from the so-called *equipartition theorem*. This is a rigorous theorem of classical statistical mechanics, which states that, in thermodynamic equilibrium at temperature T, the average energy associated with each degree of freedom of an oscillator (mode) is

$$\langle E \rangle = \frac{1}{2} k_B T \,, \tag{3.14}$$

where k_B is the Boltzmann constant. In other words, the equipartition theorem says that all modes have the same energy, irrespective of their frequencies or wavelengths.

The number of degrees of freedom is defined to be the number of squared terms appearing in the expression for the total energy of the atom (mode).

For example, consider an atom moving in three dimensions. The kinetic energy of the atom is given by

$$E = \frac{1}{2} m v_x^2 + \frac{1}{2} m v_y^2 + \frac{1}{2} m v_z^2 \,. \tag{3.15}$$

There are three quadratic terms in the energy, and therefore the atom has three degrees of freedom and a thermal energy $\frac{3}{2} k_B T$.

The energy of a single radiation mode inside a box is the energy of an electromagnetic wave

$$H = \frac{1}{2} \int_V dV \left\{ \varepsilon_0 |\vec{E}|^2 + \frac{1}{\mu_0} |\vec{B}|^2 \right\} \,. \tag{3.16}$$

Because this expression contains two squared terms, Rayleigh and Jeans argued that each mode had two degrees of freedom and therefore

$$\langle E \rangle = k_B T \,. \tag{3.17}$$

Hence, the energy density distribution (spectrum) of the blackbody radiation is given by

$$I(\lambda) = \frac{8\pi}{\lambda^4} k_B T \,. \tag{3.18}$$

This is the Rayleigh–Jeans formula, which shows that the energy density distribution depends linearly on the temperature and is inversely proportional to the fourth power of the radiation wavelength. It predicts that at any temperature, $I(\lambda)$ tends to infinity as $\lambda \to 0$. It is obviously an impossible law, since it predicts an infinite intensity at any finite temperature.

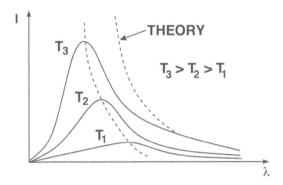

Figure 3.3 Energy density distribution (energy spectrum) of the blackbody radiation. The dashed line shows the prediction of the Rayleigh–Jeans theory. The dotted line shows the shift of the maximum of $I(\lambda)$ with temperature T.

3.3 Rayleigh–Jeans Formula versus the Experiment

Figure 3.3 shows the experimentally observed spectrum of the blackbody radiation. In this figure, $T_3 > T_2 > T_1$. Note that the maxima of the spectral lines move toward shorter wavelengths as the temperature increases. This could be expected as a radiating body changes color when the temperature is raised. However, the most interesting significance, seen in Fig. 3.3, is that the Rayleigh–Jeans formula agrees quite well with the experiment in the long-wavelength region but disagrees violently at short wavelengths.

The experimentally observed behavior shows that for some reason, the short-wavelength modes do not contribute, i.e., they are frozen out. As λ tends to zero, $I(\lambda)$ tends to zero. The theoretical Rayleigh–Jeans formula goes to infinity as λ tends to zero, leading to an absurd result known as the *ultraviolet catastrophe*. Moreover, the theoretical prediction does not even pass through a maximum.

An aside: The Rayleigh–Jeans formula represents something worse than a disagreement between theory and experiment, namely, a contradiction within the theory itself. Perfectly logical application of well-established theory led in this case to results that were not just wrong but entirely unthinkable.

Worked Example

Show that the number of modes per unit wavelength and per unit length for a string of length L is given by

$$\frac{1}{L}\left|\frac{dN}{d\lambda}\right| = \frac{2}{\lambda^2}.$$

Solution

In one dimension, volume occupied by a single mode is

$$V_k = \frac{\pi}{L}.$$

Number of modes in the volume $V = k$ is

$$N(k) = \frac{k}{V_k} = \frac{2\pi}{\lambda}\frac{L}{\pi} = \frac{2L}{\lambda}.$$

Then, the number of modes per unit wavelength and per unit length is

$$N = \frac{1}{L}\left|\frac{dN}{d\lambda}\right| = \frac{1}{L}\left|-\frac{2L}{\lambda^2}\right| = \frac{2}{\lambda^2}.$$

We can summarize the chapter on the wave theory of radiation as follows:

The experimental results on the spectrum of X-rays, properties of photoelectric effect, Compton scattering, atomic spectra, and the spectrum of the blackbody radiation indicate that something is seriously wrong with the wave theory of light!

In the next chapter, we will show how the lack of any concepts in classical physics that could explain these results led to new ideas in physics. We shall see that these results can be explained by abandoning the classical physics concepts and introducing a new concept: the particle concept of radiation. In other words, these experiments led to the failure of classical physics and the *birth* of quantum physics.

Tutorial Problem

Problem 3.1 We have shown in the chapter that the number of modes in the unit volume and the unit of frequency is

$$N = N_\nu = \frac{1}{V}\frac{dN(k)}{d\nu} = \frac{8\pi\nu^2}{c^3}.$$

In terms of the wavelength λ, we have shown that the number of modes in the unit volume and the unit of wavelength is

$$N = N_\lambda = \frac{8\pi}{\lambda^4}.$$

Explain, why it is not possible to obtain N_λ from N_ν simply by using the relation $\nu = c/\lambda$.

Chapter 4

Planck's Quantum Hypothesis: Birth of Quantum Theory

Shortly after the derivation of the Rayleigh–Jeans formula, Max Planck found a simple way to explain the experimental behavior, but in doing so he contradicted the prediction of the wave theory of light, which had been so carefully developed over the previous hundred years, that energy of an electromagnetic wave is proportional to its amplitude. Planck realized that the ultraviolet catastrophe could be eliminated by assuming that the average energy $\langle E \rangle$ depends on the frequency of a mode and each mode could only take up energy in well-defined *discrete* portions (small packets or quanta), each having the energy[a]

$$E = h\nu = \hbar\omega, \qquad \left(\hbar = \frac{h}{2\pi}, \quad \omega = 2\pi\nu \right), \qquad (4.1)$$

where the constant h is adjusted to fit the experimentally observed $I(\lambda)$.

The concept of quanta was considered by many physicists as a mathematical trick. In a published paper, Planck states: "We consider, however — this is the most essential point of the whole

---

[a] Planck was granted the Nobel Prize in 1918 for the concept of energy quanta.

Quantum Physics for Beginners
Zbigniew Ficek
Copyright © 2016 Pan Stanford Publishing Pte. Ltd.
ISBN 978-981-4669-38-2 (Hardcover), 978-981-4669-39-9 (eBook)
www.panstanford.com

calculations — E to be composed of a very definite number of equal parts and use thereto the constant of nature h." If there are n quanta in the radiation mode, the energy of the mode is $E_n = nh\nu$.

The view that light is composed of small packets of energy is directly opposed to the wave theory of light. More precisely, the contrast between the wave and Planck's hypothesis is that in the classical case the mode energy is continuous, i.e., can lie at any position between 0 and ∞ of the energy line, whereas in the quantum case the mode energy can only take on discrete (point) values.

The assumption of the discrete energy distribution required a modification of the equipartition theorem. Planck introduced "discrete portions" so that he might apply Boltzmann's statistical ideas to calculate the energy density distribution of the blackbody radiation.

4.1 Boltzmann Distribution Function

The solution to the blackbody problem may be developed from a calculation of the average energy of a harmonic oscillator of frequency ν in thermal equilibrium at temperature T.

The probability that at temperature T an arbitrary system, such as a radiation mode, has an energy E_n is given by the Boltzmann distribution

$$P_n = \frac{e^{-E_n/k_B T}}{\sum_n e^{-E_n/k_B T}}, \tag{4.2}$$

which assumes an exponential distribution of the energy between different modes. The detailed derivation of the Boltzmann distribution (Eq. 4.2) is given in Appendix A.

For quantized energy $E_n = n\hbar\omega$, we find that[a]

$$P_n = \frac{e^{-nx}}{\sum_{n=0}^{\infty} e^{-nx}}, \tag{4.3}$$

where $x = \hbar\omega/k_B T$ is a dimensionless parameter.

---

[a]We write the energy in terms of $\hbar\omega$ rather than $h\nu$ as in quantum physics the form $\hbar\omega$ is used more frequently.

Since the sum $\sum_{n=0}^{\infty} e^{-nx}$ is a particular example of a geometric series, and $|\exp(-nx)| < 1$ for all n's, we readily find that the sum tends to the limit[a]

$$\sum_{n=0}^{\infty} e^{-nx} = \frac{1}{1 - e^{-x}} . \tag{4.4}$$

Hence, we can write the Boltzmann distribution function (Eq. 4.2) in a simple form

$$P_n = \left(1 - e^{-x}\right) e^{-nx} . \tag{4.5}$$

This is a very simple formula, sometimes called the Planck distribution function. It tells us about the distribution of photons among different modes. We will use this formula in our calculations of the average energy $\langle E \rangle$, average number of photons $\langle n \rangle$, and higher statistical moments, e.g., $\langle n^2 \rangle$.

4.2 Planck's Formula for $I(\lambda)$

Assuming that n is a discrete variable, Planck showed that the average energy of the radiation mode is

$$\langle E \rangle = \sum_{n} E_n P_n = \left(1 - e^{-x}\right) \hbar\omega \sum_{n=0}^{\infty} n e^{-nx} . \tag{4.6}$$

Then, evaluating the sum in the aforementioned equation,[b] he found

$$\langle E \rangle = \frac{\hbar\omega}{e^x - 1} , \tag{4.7}$$

and finally

$$I(\lambda) = \frac{8\pi hc}{\lambda^5 \left(e^{hc/\lambda k_B T} - 1\right)} , \tag{4.8}$$

which is called **Planck's formula**, and the numerical constant h, known as the Planck constant, adjusted such that the energy density distribution (Eq. 4.8) agrees with the experimental results, is

$$h = 6.626 \times 10^{-34} \, [\text{J} \cdot \text{s}] = 4.14 \times 10^{-15} \, [\text{eV} \cdot \text{s}] . \tag{4.9}$$

---

[a] In statistical physics, the sum is called the partition function.
[b] The details of the evolution of the sum in Eq. (4.6) are left for the readers as a tutorial problem; see the Tutorial Problem 4.2

Remember the importance of Planck's idea of n being a discrete variable. Thus, Eqs. (3.18) and (4.8) for the radiation spectrum contrast the discrete energy distribution with the continuous.

How important for the explanation of the blackbody spectrum was to assume that n is a discrete variable? To answer this question, simply check how Planck's formula fits into the experimentally observed spectra. There are two interesting limited cases for λ: $\lambda \gg 1$ and $\lambda \ll 1$.

Look carefully at the denominator of Eq. (4.8). For long wavelengths, $(\lambda \gg 1)$, i.e., for a wavelength range such that $hc/\lambda k_B T \ll 1$, we can expand the exponent appearing in Eq. (4.8) into a Taylor series and when we neglect powers of $hc/\lambda k_B T$ higher than the first, we obtain

$$\left(e^{hc/\lambda k_B T} - 1\right) = 1 + \frac{hc}{\lambda k_B T} + \ldots - 1 \approx \frac{hc}{\lambda k_B T}. \qquad (4.10)$$

The Planck spectrum then gives simply

$$I(\lambda) = \frac{8\pi hc}{\lambda^5} \left(\frac{hc}{\lambda k_B T}\right)^{-1} = \frac{8\pi}{\lambda^4} k_B T, \qquad (4.11)$$

which is the Rayleigh–Jeans formula. Thus, for long wavelengths $(\lambda \gg 1)$, Planck's formula agrees perfectly with the equipartition theorem. We should point out here that *only* in the region of long wavelengths, Planck's formula and the Rayleigh–Jeans formula give the same result for the spectrum. Outside this region, discreteness brings about Planck's quantum corrections. It is most evident in the region of short wavelengths.

Let us see what Planck's formula predicts for short wavelengths. For short wavelengths, $\lambda \ll 1$, we can ignore 1 in the denominator of Eq. (4.8) as it is much smaller than the exponent. Then

$$I(\lambda) = \frac{8\pi hc}{\lambda^5 e^{hc/\lambda k_B T}}. \qquad (4.12)$$

As $\lambda \to 0$, the factor $e^{hc/\lambda k_B T} \to \infty$, and $\lambda^5 \to 0$. It is not clear that $I(\lambda)$ will go to zero as $\lambda \to 0$. However, $e^{a/\lambda}$ function goes to infinity faster than λ^5 goes to zero. Therefore, $I(\lambda) \to 0$ as $\lambda \to 0$, which agrees perfectly with the observed energy density spectrum (see Fig. 3.3).

4.2.1 *Wien and Stefan–Boltzmann Laws*

The experimental results show that $I(\lambda)$ passes through a maximum whose position depends on the temperature T. Planck's formula also shows that $I(\lambda)$ reaches a maximum whose the position depends on T and agrees perfectly with that predicted by the experimental results. The dependence of the position of the maximum on T is given by the **Wien displacement law**

$$\lambda_{max} T = \frac{hc}{4.9651 k_B} = \text{constant}, \tag{4.13}$$

where λ_{max} is the value of λ at which $I(\lambda)$ is maximal. The factor 4.9651 is a solution to the equation

$$e^{-x} + \frac{1}{5}x - 1 = 0. \tag{4.14}$$

The Wien law[a] says that with increasing temperature of the radiating body, the maximum of the intensity shifts toward shorter wavelengths. In terms of energy, higher temperatures tend to give higher photon energies.

Moreover, Planck's formula correctly predicts the experimental observation, the **Stefan–Boltzmann law**[b], giving the variation of the total power per unit area of the emitted blackbody radiation with the fourth power of its absolute temperature

$$I = \frac{c}{4} \int_0^\infty I(\lambda) d\lambda = \sigma T^4, \tag{4.15}$$

where I is the total intensity of the emitted radiation and σ is a constant, called the Stefan–Boltzmann constant.[c] The factor $c/4$, where c is the speed of light, arises from the relation between the intensity spectrum (radiance) and the energy density distribution. The relation follows from classical electromagnetic theory.[d]

---

[a]Wien was granted the Nobel Prize in 1911 for his discoveries regarding the laws governing the radiation of heat.

[b]The Stefan–Boltzmann law was found experimentally by Stefan and later deduced theoretically by Boltzmann.

[c]Details of the derivation of the Stefan–Boltzmann law are left as a tutorial problem; see Tutorial Problem 4.7.

[d]Apart from the factor 1/4, one could have guessed the multiplication by speed of light using the following analysis: To turn density (energy/volume) into power/area, we have to multiply by something with units of distance/time, and the only relevant speed in the problem involving radiation is the speed of light.

The Stefan–Boltzmann law is reflected in the blackbody spectrum as the areas under the intensity profile.[a] As the temperature increases, the area under the intensity profile increases rapidly with T.

An interesting observation: *Quantum against continuous*
To explore the importance of the discrete distribution of the radiation energy, in other words, the importance to assume that n is a discrete variable, it is useful to compare Planck's formula for a discrete n with that for a continuous n.

Thus, assume for a moment that n is a continuous, rather than a discrete, variable. Then the Boltzmann distribution takes the form

$$P_n = \frac{e^{-nx}}{\int\limits_0^\infty dn\, e^{-nx}} \, , \qquad (4.16)$$

and hence the average energy is given by

$$\langle E \rangle = \hbar\omega \, \frac{\int\limits_0^\infty dn\, n e^{-nx}}{\int\limits_0^\infty dn\, e^{-nx}} = -\hbar\omega \, \frac{(1/x)'}{1/x} = \frac{\hbar\omega}{x} = k_B T \, , \qquad (4.17)$$

where $'$ denotes the first derivative of $1/x$ with respect to x.

This result is the one expected from the classical equipartition theorem.

Looking backward with the knowledge of the quantum hypothesis, we see that the essence of the blackbody calculation is remarkably simple and provides a dramatic illustration of the profound difference that can arise from summing things discretely instead of continuously, i.e., making an integration.

We see from the above that the physical world prefers a sum over an integration, but why it happens still escapes our understanding.

---

[a]The thermal radiation from many real materials that are not black is found experimentally to be very nearly proportional to the fourth power of the absolute temperature, but with a proportionality constant, which is smaller than σ.

More interesting observations: *Energy and mass measured with a clock*

Time is measured by a periodically oscillating device, a clock.

Since, according to the quantum hypothesis, energy of the radiation depends on time

$$E = h\nu = \frac{h}{\tau},$$ (4.18)

where τ is the period of oscillation, it also should be measured with a clock.

Even more, since

$$E = mc^2 = \frac{h}{\tau},$$ (4.19)

mass should be measured with a clock as well.

Tutorial Problems

Problem 4.1 Consider Planck's formula for two temperatures $T_1 = 2000$ K and $T_2 = 4000$ K.

(a) How much the maximum of $I(\lambda)$ for T_2 is shifted relative to the maximum for T_1?

(b) How much the total area under the graph for T_2 is larger than that under the graph for T_1?

Problem 4.2 Using Planck's formula for P_n,

(a) Show that the average number of photons is given by

$$\langle n \rangle = \frac{1}{e^x - 1},$$

where $x = \frac{\hbar\omega}{k_B T}$, and k_B is the Boltzmann constant.

(b) Show that for large temperatures ($T \gg 1$), the average energy is proportional to temperature, i.e., $\langle E \rangle = k_B T$, that the prediction of Planck's formula agrees with the Rayleigh–Jeans formula.

(c) Calculate the variance of the number of photons defined as $\sigma_n = \langle n^2 \rangle - \langle n \rangle^2$ and show that the ratio

$$\alpha = \frac{\langle n^2 \rangle - \langle n \rangle}{\langle n \rangle^2} = 2.$$

In statistical physics, the ratio α is used as a measure of the departure of photon statistics from a Poissonian distribution: $\alpha < 1$ means sub-Poissonian, $\alpha = 1$ means Poissonian, and $\alpha > 1$ means super-Poissonian distribution of photons.

Problem 4.3 Suppose that photons in a radiation field have a Poissonian distribution defined as

$$P_n = \frac{\langle n \rangle^n}{n!} e^{-\langle n \rangle} .$$

Calculate the variance of the number of photons and show that the ratio $\alpha = 1$.

Problem 4.4 Show that the Boltzmann formula for the probability distribution, Eq. (4.5), can be written as

$$P_n = \frac{\langle n \rangle^n}{(1 + \langle n \rangle)^{n+1}} ,$$

where

$$\langle n \rangle = \frac{1}{e^x - 1}$$

is the average number of photons of frequency ω and $x = \hbar \omega / k_B T$.

Problem 4.5 Show that at the wavelength λ_{max}, the intensity $I(\lambda)$ calculated from Planck's formula has its maximum

$$I(\lambda_{max}) \approx \frac{170 \pi (k_B T)^5}{(hc)^4} .$$

Problem 4.6 *Wien displacement law*

(a) Derive the Wien displacement law by solving the equation $dI(\lambda)/d\lambda = 0$.

(*Hint:* Set $hc/\lambda k_B T = x$ and show that dI/dx leads to the equation $e^{-x} = 1 - \frac{1}{5}x$. Then show that $x = 4.9651$ is the solution).

(b) In part (a), we have obtained λ_{max} by setting $dI(\lambda)/d\lambda = 0$. Transform Planck's formula from the wavelength to frequency dependence and calculate ν_{max} by setting $dI(\nu)/d\nu = 0$. Is it possible to obtain ν_{max} from λ_{max} simply by using $\lambda_{max} = c/\nu_{max}$? Note, ν_{max} is the frequency at which the intensity of the emitted radiation is maximal.

Problem 4.7 *Derivation of the Stefan–Boltzmann law*
Derive the Stefan–Boltzmann law evaluating the integral in Eq. (4.15).

Hint: It is convenient to evaluate the integral by introducing a dimensionless variable

$$x = \frac{hc}{\lambda k_B T} \, .$$

4.3 Photoelectric Effect: Quantum Explanation

We have just learned about the spectacular triumph of quantum hypothesis to be able to explain the experimentally observed blackbody radiation spectra. The quantum hypothesis of light was also successful in explaining the photoelectric effect. It was explored by Albert Einstein[a] who extended Planck's hypothesis by postulating that these discrete quanta of energy $h\nu$, which are emitted at time, can also propagate as individual well-localized quanta (particles of light), which he termed *photons*.[b]

Einstein proposed that the photoelectric effect can be explained in terms of the energy conservation. The energy of a single photon is $E = h\nu$, and then the photoelectric effect is given by the energy conservation formula

$$h\nu = W + \frac{1}{2}mv_{max}^2 \, , \qquad (4.20)$$

where W is the work function required to remove an electron from the plate, and v_{max} is the maximal velocity of the removed electrons.

The photoelectric effect showed that photons of an incident light of a frequency less than a threshold frequency ν_T (a cutoff frequency) do not have enough energy to remove an electron from a particular plate. From Einstein's photoelectric formula, we see that the minimum energy required to remove an electron from the plate is

---

[a]Einstein was granted the Nobel Prize in 1921 for his discovery of the law of photoelectric effect.

[b]In fact, the term "photon" was first introduced by Gilbert Lewis in 1926 (Nature **118**, 784 (1926)). The term "photon" did not appear in Einstein's paper on the photoelectric effect. He used the term "localized quanta of energy."

$$hv_T = W,\qquad\qquad(4.21)$$

which agrees with the experimental prediction.

The stopping potential—the potential at which the photoelectric current does drop to zero—is found from

$$eV_s = \frac{1}{2}mv_{max}^2,\qquad\qquad(4.22)$$

which gives

$$V_s = \frac{\frac{1}{2}mv_{max}^2}{e} = \frac{hv - W}{e}.\qquad\qquad(4.23)$$

Thus, we see that the stopping potential is proportional to frequency and increases with increasing v, which is in accordance with the experimental observation.

We may conclude that what we have just shown is an another example of a remarkable triumph of Planck's quantum hypothesis. Einstein's photoelectric formula (4.20), incorporating the photon nature of light, correctly explains the properties of the photoelectric effect discovered by Hertz. It predicts correctly that the photoelectric effect depends on the frequency of the incident light, not upon its intensity, contrary to what the wave theory suggested.

4.4 Compton Scattering: Quantum Explanation

Another support of Planck's hypothesis, which gave the most conclusive confirmation for the corpuscular (photon) aspect of light, was provided by the Compton scattering effect.

Let us re-examine the Compton scattering experiment, discussed in Section 2.3, but now we shall assume that the incident X-rays are composed of particles (photons) that can be scattered from the electrons as balls. Suppose that the incident photon has momentum p and energy $E = pc$. The scattered photon has momentum p' and energy $E' = p'c$. The electron is initially at rest, so its energy is $E_e = m_0c^2$, and the initial momentum is zero.

From the energy conservation, we have

$$E + E_e = E' + mc^2,\qquad\qquad(4.24)$$

where m is the relativistic mass

$$m = \frac{m_0}{\sqrt{1 - (v/c)^2}} \, , \tag{4.25}$$

and v is the velocity of the scattered electron.

Since $E = pc$, $E' = p'c$, and $E_e = m_0 c^2$, the energy conservation formula may be written as

$$\left(pc - p'c + m_0 c^2 \right) = mc^2 \, . \tag{4.26}$$

Taking square of both sides of Eq. (4.26), we obtain

$$\left(pc - p'c + m_0 c^2 \right)^2 = (mc^2)^2 = (m_0 c^2)^2 + (p_e c)^2 \, , \tag{4.27}$$

where p_e is the momentum of the scattered electron. Thus, we can write

$$\left(p - p' \right)^2 + 2m_0 c \left(p - p' \right) = p_e^2 \, . \tag{4.28}$$

This energy conservation formula contains the momentum of the scattered electron. Unfortunately, the momentum of the scattered electrons is difficult to measure in experiments, as they move with large velocities. Therefore, we will eliminate p_e using the momentum conservation law:

$$\vec{p}_e = \vec{p} - \vec{p}' \, , \tag{4.29}$$

from which we find

$$p_e^2 = \vec{p}_e \cdot \vec{p}_e = p^2 + p'^2 - 2pp' \cos \alpha \, , \tag{4.30}$$

where α is the angle between directions of the incident and scattered photons.

Substituting Eq. (4.30) into Eq. (4.28), we get

$$2m_0 c \left(p - p' \right) = 2pp' (1 - \cos \alpha) \, , \tag{4.31}$$

from which, we find the momentum difference

$$p - p' = \frac{pp'}{m_0 c} (1 - \cos \alpha) \, . \tag{4.32}$$

This is how far we can get with the analysis of the Compton effect using arguments of classical physics. To relate this effect to the wavelength of the radiation, we apply Planck's hypothesis of the quantization of energy. In this case, we find the following relation between the momentum and wavelength:

$$p = \frac{E}{c} = \frac{h\nu}{c} = \frac{h}{\lambda} \, . \tag{4.33}$$

Using this relation in Eq. (4.32), we finally obtain

$$\lambda' - \lambda = \frac{h}{m_0 c} (1 - \cos\alpha) \, . \qquad (4.34)$$

This is the Compton scattering formula.

The quantity $h/(m_0 c)$ is called *the Compton wavelength* of a particle with rest mass m_0 (here, an electron)

$$\lambda_c = \frac{h}{m_0 c} = 2.426 \times 10^{-12} \quad [\text{m}] \, . \qquad (4.35)$$

The Compton scattering formula (Eq. 4.34) shows that the change in wavelength depends *only* on the scattering angle α. During the collision, the photon loses a part of its energy and consequently its wavelength increases. Thus, the quantum theory predicts correctly that the scattered light has different wavelength than the incident light. The classical (wave) theory predicts that $\lambda' = \lambda$.

One may notice from the Compton scattering formula (Eq. 4.34) that the transition from quantum (photon) to classical (wave) description of the theory is to put $h \to 0$. Thus, the Planck constant distinguishes between quantum and classical, but the presence of the constant h is not the universal criterion for quantum.

An interesting observation

A significant feature of the derivation of the Compton scattering formula is that it relied essentially on special relativity. Thus, the Compton effect not only confirms the existence of photons, it also provides a convincing proof of the validity of special relativity.

Worked Example

A photon collides with a stationary electron.

(a) Show that in the collision the photon cannot transfer all its energy to the electron.
(b) Show that a photon cannot produce a positron–electron pair.

Solution

(a) Assume that the photon can transfer all its energy to the electron. Then, from the conservation of energy

$$E_f + m_0c^2 = mc^2 = \sqrt{m_0^2c^4 + p_e^2c^2} , \qquad (4.36)$$

where $E_f = h\nu$ is the energy of the photon, and p_e is the momentum of the electron. From the conservation of momentum, we have

$$p_f = \frac{h\nu}{c} = p_e , \qquad (4.37)$$

and substituting Eq. (4.37) into Eq. (4.36), we obtain

$$h\nu + m_0c^2 = \sqrt{m_0^2c^4 + (h\nu)^2} , \qquad (4.38)$$

which is not true, as the right-hand side is larger than the left-hand side.

(b) From the conservation of momentum, we have

$$p_f = \sqrt{m_0^2c^2 + p_e^2} + \sqrt{m_0^2c^2 + p_p^2} , \qquad (4.39)$$

where p_p is the momentum of the positron.
We see from the above equation that

$$p_f > |\vec{p}_e| + |\vec{p}_p| \geq |\vec{p}_e + \vec{p}_p| . \qquad (4.40)$$

Hence

$$\vec{p}_f > \vec{p}_e + \vec{p}_p . \qquad (4.41)$$

Thus, a photon cannot produce a positron–electron pair.

Tutorial Problems

Problem 4.8 Explain why is it much more difficult to observe the Compton effect in the scattering of visible light than in the scattering of X-rays?

Problem 4.9 Calculate the kinetic energy transferred to a stationary proton hit by a photon of energy $h\nu$.

Problem 4.10 Consider the Compton scattering.

(a) Show that $\Delta E/E$, the fractional change in photon energy in the Compton effect, equals

$$\frac{\Delta E}{E} = \frac{h\nu'}{m_0 c^2} (1 - \cos\alpha) ,$$

where ν' is the frequency of the scattered photon and $\Delta E = E - E'$.

(b) Show that the relation between the directions of motion of the scattered photon and the recoil electron in the Compton scattering is

$$\cot\frac{\alpha}{2} = \left(1 + \frac{h\nu}{m_0 c^2}\right) \tan\theta ,$$

where α is the angle of the scattered photon, θ is the angle of the recoil electron, and ν is the frequency of the incident light.

This formula shows that one can test the Compton effect by a measurement of the angles α and θ instead of measuring the wavelength λ' of the scattered photons, which is difficult to measure with a large precision in experiments.

Chapter 5

Bohr Model

Anyone who is not shocked by quantum theory has not
understood it.

—Niels Bohr

In 1913, the Danish scientist Niels Bohr used Planck's concept to propose a model of the hydrogen atom[a] that had a spectacular success in explanation of the discrete atomic spectra.[b] The model also correctly predicted the wavelengths of the spectral lines. We have seen that the atomic spectra exhibit discrete lines unique to each atom. From this observation, Bohr concluded that atomic electrons can have only certain discrete energies. That is, the kinetic and potential energies of electrons are limited to only discrete particular values, as the energies of photons in a blackbody radiation.

---

[a]The Bohr model is called "semiclassical" not "quantum" because, as we shall see, it contains concepts of both classical and quantum physics.

[b]About the earlier models of the hydrogen atom (Thomson model, Rutherford model), see R.A. Serway, C.J. Moses, and C.A. Moyer, *Modern Physics* (Saunders, New York, 1989).

Quantum Physics for Beginners
Zbigniew Ficek
Copyright © 2016 Pan Stanford Publishing Pte. Ltd.
ISBN 978-981-4669-38-2 (Hardcover), 978-981-4669-39-9 (eBook)
www.panstanford.com

5.1 Hydrogen Atom

In the formulation of his model, Bohr assumed that the electron in the hydrogen atom moves under the influence of the Coulomb attraction between it and the positively charged nucleus, as assumed in classical mechanics. However, he incorporated new ideas. Namely, he *postulated* that the electron could only move in certain nonradiating orbits, which he called *stationary orbits* (stationary states). Next, he postulated that the atom radiates only when the electron makes a transition between states.

Let us illustrate Bohr's ideas in some detail. We begin with considering the classical equation of motion for the electron in a circular orbit, which is based on Newton's laws of motion and Coulomb's law of electric force. The electrostatic Coulomb force provides the centripetal acceleration v^2/r that holds the electron in an orbit r from the nucleus. The condition for orbit stability is

$$\frac{e^2}{4\pi\varepsilon_0 r^2} = m_e\frac{v^2}{r}. \tag{5.1}$$

This relation allows us to calculate kinetic energy of the electron

$$K = \frac{1}{2}m_e v^2 = \frac{e^2}{8\pi\varepsilon_0 r}, \tag{5.2}$$

which, together with the potential energy

$$U = -\frac{e^2}{4\pi\varepsilon_0 r} \tag{5.3}$$

gives the total energy of the electron as

$$E = K + U = -\frac{e^2}{8\pi\varepsilon_0 r}. \tag{5.4}$$

From the kinetic energy, we can find the velocity of the electron, its linear momentum, and finally its angular momentum

$$L = m_e vr = \sqrt{\frac{m_e e^2 r}{4\pi\varepsilon_0}}. \tag{5.5}$$

Up to this point, the analysis has been classical and as such creates a problem. The equations (5.4) and (5.5) show that the energy and angular momentum of the electron depend on the radius of the orbit. An obvious question arises: How to find the radius, since in practice it is rather impossible to measure r?

5.1.1 Quantization of Angular Momentum

Bohr got the tricky idea to solve this problem. To find the radius of the orbit, Bohr *postulated* that the **angular momentum of the electron is quantized**, i.e., it can only take values that are integer multiples of $\hbar$.

Where this idea came from?

It came from the following observation: One can notice from Planck's formula $E = h\nu$ that h has the units of energy multiplied by time (J·s), or equivalently of momentum multiplied by distance. The electron in the atom travels a distance $2\pi r$ per one turn. Since the momentum is $p = m_e v$, we obtain

$$(m_e v)(2\pi r) = nh, \tag{5.6}$$

which shows that the angular momentum is quantized:

$$L = n\hbar, \qquad n = 1, 2, 3, \ldots . \tag{5.7}$$

Comparing this quantum relation for the angular momentum with Eq. (5.5), we can readily find the radius of the orbits. Since

$$\frac{m_e e^2 r}{4\pi \varepsilon_0} = n^2 \frac{h^2}{4\pi^2}, \tag{5.8}$$

we find, as a result of the quantization of angular momentum, that the orbits are quantized:

$$r = n^2 \frac{\varepsilon_0 h^2}{\pi m_e e^2} = n^2 a_o, \tag{5.9}$$

where

$$a_o = \frac{\varepsilon_0 h^2}{\pi m_e e^2} = 5.3 \times 10^{-11} \, [\text{m}] \tag{5.10}$$

is called the Bohr radius. It is the radius of the electron's lowest energy level and is taken as the length scale in the quantum description of the hydrogen atom. The result (5.9) is in terms of known constants and is very different from what one could expect from classical physics. The electron's orbits cannot have any radius; only certain radii are allowed. The radius of the electron's orbit may be a_o, $4a_o$, $9a_o$, $\ldots$, but *never* $2a_o$ or $1.6a_o$.

By substituting Eq. (5.9) into Eq. (5.4), we find that the energy of the electron is also quantized:

$$E_n = -\frac{1}{(4\pi \varepsilon_0)^2} \frac{m_e e^4}{2\hbar^2} \frac{1}{n^2}. \tag{5.11}$$

The energies specified by Eq. (5.11) are called the *energy levels* or energy states of the hydrogen atom. These levels are all negative, signifying that the electron does not have enough energy to escape from the atom. This is the so-called *bounding energy*. The lowest energy level $n = 1$ is called the ground state of the atom and, in the hydrogen atom, has the energy

$$E_1 = -\frac{m_e e^4}{8\varepsilon_0^2 h^2} = -13.6 \quad [\text{eV}]. \tag{5.12}$$

This is the energy required to separate a hydrogen atom into a proton and an electron.

The higher levels $n = 2, 3, \ldots$ are called excited states. Note also that the energy becomes less negative as n increases. At $n \to \infty$, $E_n \to 0$ and there is no binding energy of the electron to the nucleus. We say the atom is ionized.

5.1.2 Quantum Jumps Hypothesis

How does the electron make transitions between the energy levels?

Bohr introduced another postulate—the hypothesis of *quantum jumps*—that the electron jumps (suddenly moves) from one energy level to another emitting or absorbing radiation of a frequency

$$\nu = \frac{E_m - E_n}{h} = \frac{m_e e^4}{8\varepsilon_0^2 h^3} \left(\frac{1}{n^2} - \frac{1}{m^2} \right), \tag{5.13}$$

or of wavelength

$$\lambda = \left(\frac{m_e e^4}{8c\varepsilon_0^2 h^3} \right)^{-1} \left(\frac{1}{n^2} - \frac{1}{m^2} \right)^{-1} = R^{-1} \left(\frac{1}{n^2} - \frac{1}{m^2} \right)^{-1}, \tag{5.14}$$

where R is the Rydberg constant.

Equation (5.14) states that the radiation emitted by excited hydrogen atoms should contain certain wavelengths only. The wavelengths fall into definite sequences, called *spectral series*, that depend upon the quantum number of the final energy level. This is illustrated in Fig. 5.1. Note that the energy levels get closer together (converge) as the n value increases.

We should point out here that quantum jumps are inconsistent with any classical picture of radiation and are often regarded as a *quantum effect without classical analog.*

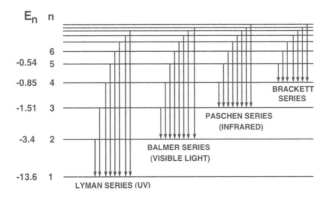

Figure 5.1 Energy-level diagram for hydrogen atom with possible electron transitions.

In summary, the Bohr model was very successful in explaining the discrete atomic spectra of one-electron (hydrogen-like) atoms. Simply, discrete spectral lines are observed because in order for an electron to move from one state to another, it must either absorb or emit exactly the right amount of energy to jump. This puts the model among the most spectacular achievements in physics.[a]

5.2 Franck–Hertz Experiment

The Bohr model is, of course, a theoretical model of an atom. It predicts that the electron in the atom can have only certain discretely separated energy levels. At this point, we may ask a question: Do in reality the quantized energy levels really exist? The observation of the discrete atomic spectra is one of the evidences of discretely separated energy levels. However, atomic spectra are not the only means of investigating the presence of discrete energy levels.

In 1914, shortly after the Bohr model was introduced, James Franck and Gustav Hertz performed a series of experiments that provided a direct demonstration that atomic energy levels *do* indeed

---

[a]Bohr was granted the Nobel Prize in 1922 for investigating the structure of atoms and of the radiation emanating from them.

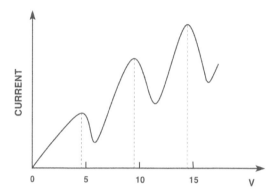

Figure 5.2 Results of the Franck–Hertz experiment for mercury atoms showing regular maxima and minima of the anode current.

exist. In the experiments, electrons emitted by a heated cathode were accelerated by a potential V applied between the cathode and anode plate placed in a tube filled with a vapor of Hg atoms.

The measurements involved measuring the anode current as a function of the voltage V. The results of the experiment for the tube containing mercury (Hg) vapor are shown in Fig. 5.2. As the accelerating potential is increased, the anode current is increased as more electrons arrive at the anode plate. It indicates that the presence of the mercury atoms does not affect the motion of the electrons and that no energy is transferred from the electrons to the atoms. However, when V reaches 4.9 V, the current abruptly drops. This can be interpreted that some interaction between the electrons and the Hg atoms suddenly begins when the electrons attain a kinetic energy 4.9 eV. Additional drops of the current are observed at integer multiples of 4.9 eV.

To understand why the anode current drastically drops at integer multiples of 4.9 eV, Franck and Hertz,[a] in addition to the current, observed the emission (fluorescence) spectrum of the Hg vapor. They observed that the wavelength of the emitted radiation was 2536 Å, which for the Hg atoms corresponds exactly to a photon energy of 4.9 eV. Thus, in a single collision between the electrons

---

[a]Franck and Hertz were granted the Nobel Prize in 1925 for their discovery of the laws governing the impact of an electron on an atom.

and the atoms, the energy of 4.9 eV is transferred from the electrons to the atoms. The current drops at 9.8 eV, 14.7 eV, etc. correspond to multiple, two, three, etc., collisions, respectively.

To summarize, the Franck–Hertz experiment provided a proof that atomic energy states are quantized. It also provided a method for the direct measurement of the energy difference between the quantum energy states of an atom.

5.3 X-Rays Characteristic Spectra

In the study of X-rays characteristic spectra, we observed that the spectra were composed of two distinctive lines (see Fig. 2.1) whose origin was not accountable in terms of classical electromagnetic theory.

In 1913, Henry Moseley studied X-rays characteristic spectra in detail, and he showed how the X-ray spectra can be understood in terms of the quantum theory of radiation and on the basis of the discrete energy levels of atoms in the anode material. His analysis was based on the Bohr model, and the explanation was as follows:

In a multi-electron atom, the fast and energetic electrons from the cathode knock electrons of the anode atoms out of their inner orbits, as shown in Fig. 5.3. Then, the outer electrons jump to these

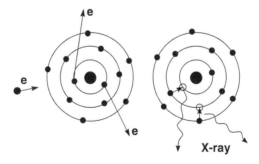

Figure 5.3 Illustration of the X-ray emission from a multi-electron atom.

empty places emitting X-ray (short-wavelength) photons of discrete frequencies.

It is interesting to note that generation of X-rays is often called the inverse photoelectric effect. Energy of moving electrons is converted into photons.

5.4 Difficulties of Bohr Model

Bohr was able to predict the stationary properties of the hydrogen-like atom, i.e., the energy levels with great accuracy, but the dynamics had to be introduced artificially, by introducing the concept of quantum (sudden) jumps between the energy states. This ad hoc assumption, which has no counterpart in classical physics, caused a vivid controversy in quantum physics. In addition, there were many objections to the Bohr theory, and to complete our discussion of this theory, we indicate some of its undesirable aspects:

- The model contains both the classical (orbital) and quantum (jumps) concepts of motion.
- The model was applied with a mixed success to the structure of atoms more complex than hydrogen.
- Classical physics does not predict the circular Bohr orbits to be stable. An electron in a circular orbit is accelerating toward the center and, according to classical electrodynamic theory, should gradually lose energy by radiation and spiral into the nucleus.
- The model does not tell us how to calculate the intensities of the spectral lines.
- If the electron can have only particular energies, what happens to the energy when the electron jumps from one orbit to another?
- How the electron knows that it can jump only if the energy supplied is equal to $E_m - E_n$?
- The model does not explain how atoms can form different molecules.
- Experiments showed that some of the lines in atomic spectra are not singlets but are composed of two or more closely

spaced lines. The lines can be resolved by applying a magnetic field (Zeeman effect), the feature not presented in the Bohr model.

We see that some of these objections are really of a very fundamental nature, and much effort was expended in attempts to develop a quantum theory that would be free of these objections. As we will see later, the effort was well rewarded and led to what we now know as quantum wave mechanics. Nevertheless, the Bohr theory is still frequently employed as the first approximation to the more accurate description of quantum effects. In addition, the Bohr theory is often helpful in visualizing processes that are difficult to visualize in terms of the rather abstract language of the quantum wave mechanics, which will be presented in details in next few chapters.

Tutorial Problems

Problem 5.1 We usually visualize electrons and protons as spinning balls. Is it a true model? To answer this question, consider the following example.

Suppose that the electron is represented by a spinning ball. Using Bohr's quantization postulate, find the linear velocity of the electron's sphere. Assume that the radius of the electron is in the order of the radius of a nucleus, $r \approx 10^{-15}$ m (1 fm). What would you say about the validity of the spinning ball model of the electron?

Problem 5.2 *How big is an electron?*
Suppose that the electron is a spherical shell of radius r_e and all the electron's charge is evenly distributed on the shell. Using the formula for the energy of a charged shell, calculate the classical electron radius. Compare the size of the electron with the size of an atomic nucleus.

This problem illustrates the discrepancy between the experimentally observed size of the electron and that calculated from the classical electromagnetic theory.

Problem 5.3 Just for fun with Maxwell's equations. A spatial distribution of charges creates the radial electric field of the form

$$\vec{E}(r) = \frac{Ae^{-br}}{r^2}\,\hat{r}\,,$$

where A and b are real and positive constants and $\hat{r}$ is the unit vector in the radial direction.

Using the Maxwell equation (1.1), calculate the density of the charge and show that the field is produced by a positive charge located at the origin, $r = 0$, and a negative charge continuously distributed in space. Then, show that the total charge is zero.

This problem illustrates Rutherford's (classical) model of an atom, used before the Bohr model was introduced. The positive charge at the origin represents the charge of the nucleus and the negative charge symmetrically distributed in space represents the cloud of electrons.

Problem 5.4 *The Bohr model for a hydrogen-like atom*
Show that in the Bohr atom model, the electron's orbits in a hydrogen-like atom are quantized with the radius $r = n^2 a_0/Z$, where $a_0 = 4\pi\varepsilon_0\hbar^2/me^2$ is the Bohr radius, $n = 1, 2, \ldots$, and Z is atomic number. $Z = 1$ refers to a hydrogen atom, $Z = 2$ to a Helium He^+ ion, and so on.

Problem 5.5 *Magnetic dipole moment of an electron*
The magnetic dipole moment $\vec{\mu}$ of a current loop is defined by $\vec{\mu} = I\vec{S}$, where I is the current and $\vec{S} = S\vec{u}$ is the area of the loop, with $\vec{u}$— the unit vector normal to the plane of the loop. A current loop may be represented by a charge e rotating at constant speed in a circular orbit. Use the classical model of the orbital motion of the electron and Bohr's quantization postulate to show that the magnetic dipole moment of the loop is quantized such that

$$\mu = n\,m_B\,, \qquad n = 1, 2, 3, \ldots\,,$$

where $m_B = e\hbar/2m_e$ is the Bohr magneton and m_e is the mass of the electron.

Problem 5.6 *Why we do not see single photons?*
Consider an experiment. A student is at a distance 10 m from a light source whose power $P = 40$ W.

(a) How many photons strike the student's eye if the wavelength of the light is 589 nm (yellow light) and the radius of the pupil (a variable aperture through which light enters the eye) is 2 mm.

(b) At what distance from the source only one photon would strike the student's eye?

Problem 5.7 *Stern–Gerlach experiment: Evidence of the quantized angular momentum (spin) of an electron*

(a) Illustrate and explain in a simple way the Stern–Gerlach experiment.

(b) Explain, using some algebra, why in the Stern–Gerlach experiment the silver atomic beam after passing the magnetic field is not continuously spread, but is split into only two components?

Challenging Problem: *Collapse of the classical atom*

The classical atom has a stability problem. Let us model the hydrogen atom as a non-relativistic electron in a classical circular orbit about a proton. From the electromagnetic theory, we know that a deaccelerating charge radiates energy. The power radiated during the deacceleration is given by the Larmor formula (2.3).

(a) Show that the energy lost *per cycle* is small compared to the electron's kinetic energy. Hence, it is an excellent approximation to regard the orbit as circular at any instant, even though the electron eventually spirals into the proton.

(b) How long does it take for the initial radius of $r_0 = 1$ Å to be reduced to zero? Insert appropriate numerical values for all quantities and estimate the (classical) lifetime of the hydrogen atom.

Chapter 6

Duality of Light and Matter

If all this damned quantum jumps were really to stay, I should be sorry I ever got involved with quantum theory.

—Erwin Schrödinger

We have already encountered several aspects of quantum physics, but in all the discussions so far, we have always assumed that a particle, a photon in particular, is a small solid object. However, quantum physics as it developed in the three decades after Planck's discovery, found a need for an uncomfortable fusion of the discrete and the continuous. This applies not only to light but also to particles. Arguments about particles or waves gave way to a recognized need for both particles and waves in the description of radiation. Thus, we will see that our modern view of the "true nature" of radiation and matter is that they have a dual character, behaving like a wave under some circumstances and like a particle under other, but never both simultaneously.

In the last few chapters, we discussed the wave and particle properties of light, and with our current knowledge of the radiation theory, we can recognize the following wave and particle aspects of

Quantum Physics for Beginners
Zbigniew Ficek
Copyright © 2016 Pan Stanford Publishing Pte. Ltd.
ISBN 978-981-4669-38-2 (Hardcover), 978-981-4669-39-9 (eBook)
www.panstanford.com

radiation:

Wave character	Particle character
1. Polarization	Photoelectric effect
2. Interference	Compton scattering
3. Diffraction	Blackbody radiation

The fact that the same light beam can behave like a wave and a particle raises tricky questions.

How can light be a wave and a particle at the same moment?

Is a photon a particle or a wave?

Evidently, we are left with an obvious question: Which theory are we to believe?

If we accept the duality of light, then another obvious question arises: Is this dual character a property of light alone or of all structures in the universe, in particular material particles?

Following this, one may ask: Is an electron really a material particle or is it a wave?

We have to say that there is no definite answer to these questions. We can get a satisfactory answer to these questions, but we must enter the strange and often deceiving world of **quantum wave physics**.

6.1 Matter Waves

The confusion over particle versus wave properties was resolved by *Louis de Broglie*. On the basis of the observations that

- Nature is strikingly symmetrical and
- Our universe is composed entirely of light and matter,

he postulated that since light has a dual wave–particle nature, matter also has this nature.

The dual nature of light shows up in equations

$$\lambda = \frac{h}{p}, \qquad E = h\nu. \tag{6.1}$$

Each equation contains within its structure both a wave concept (λ, ν), and a particle (p, E). According to this observation, a particle with energy E and momentum p should have the possibility to manifest itself as a wave.

The photon also has an energy given by the relationship from the relativity theory

$$E = mc^2 . \tag{6.2}$$

Since $E = h\nu = hc/\lambda$, we find the wavelength

$$\lambda = \frac{h}{mc} = \frac{h}{p} , \tag{6.3}$$

where p is the momentum of the photon.

This does not mean that light has mass, but because mass and energy can be interconverted, it has an energy that is equivalent to some mass.

De Broglie postulated that a particle can have a wave character and predicted that the wavelength of a matter wave would also be given by the same equation that held for light, where now p would be the momentum of the particle[a]

$$\lambda = \frac{h}{mv} = \frac{h}{p} , \tag{6.4}$$

where v is the velocity of the particle, and λ is called the de Broglie wavelength.

Remember this formula! It is the fundamental matter–wave postulate and will appear very often in our journey through the developments of quantum physics.

If particles may behave as waves, could we ever observe the matter waves?

The idea was to perform a diffraction experiment with electrons. But an obvious question was: How to perform such an experiment? What wavelengths can we expect? To answer these questions and gain some appreciation of the de Broglie wavelength, consider first a simple example.

---

[a]Wave character of particles is often interpreted that particles itself are waves. It is a wrong interpretation! Quantum physics predicts that particles can *behave* as waves, and it does not mean that they are waves in nature.

Example: What is the de Broglie wavelength of an electron whose kinetic energy is $K = 100$ eV?

We first calculate the velocity of the electron from which we then find the de Broglie wavelength corresponding to that velocity.

Thus, the velocity of the electron of energy 100 eV is

$$v = \sqrt{\frac{2K}{m_e}} = 5.9 \times 10^3 \quad [\text{km/s}].$$

Hence, the de Broglie wavelength corresponding to this velocity is

$$\lambda = \frac{h}{p} = \frac{h}{m_e v} = 1.2 \quad [\text{Å}].$$

The wavelength is very short; it is about the size of a typical atom.

The aforementioned example shows that the wavelength is very short. Thus, an obvious question arises: How to detect such short wavelengths?

We may notice from the aforementioned example that the wavelength is also of the same order as the wavelengths of X-rays. Therefore, we immediately conclude that the matter waves can be detected in the same way that the wave nature of X-rays was first observed: diffraction of particles on crystals.

This idea was tested experimentally in 1926 by Clinton Davisson and Lester Germer,[a] and independently by George Thompson,[b] who performed electron-scattering experiments. They observed that electrons, after passing through a large nickel crystal ($d = 2.15$ Å), produced an interference pattern. Using the experimental data, they found that the wavelength calculated from the diffraction relation

$$n\lambda = 2d \sin \theta_n, \qquad n = 0, 1, 2, \ldots \tag{6.5}$$

was in excellent agreement with the wavelength calculated from the de Broglie relation $\lambda = h/p$.

An interference pattern was demonstrated not only with electrons, but also with many kinds of material particles like protons and neutrons. However, any attempt to observe an interference

---

[a] In fact, Davisson and Germer had done some of these experiments even before de Broglie's matter wave postulate was established, but they could not interpret their results.

[b] Davisson and Thompson were granted the Nobel Prize in 1937 for their experimental discovery of the diffraction of electrons by crystals.

pattern with neutral atoms was found difficult by the fact that atoms carried no charge as electrons do and could not penetrate through condensate matter like neutrons.

This problem has not been solved until recently, when rapidly developing laser techniques enabled to create two slits separated by a very small distance. In addition, laser cooling of atoms has enabled to increase the de Broglie wavelength of an atom, thereby allowing to observe interference effects in Young's double-slit type experiment with atoms traveling along well-separated paths.

In 1991, Carnal and Mlynek demonstrated that atoms emerging from the same source and split by two slits produced an interference pattern at the atomic detector. The source of the atoms in their experiment was a thermal beam of metastable helium atoms. The velocity of the atoms was adjusted by setting the temperature of the source to $T = 295$ K, corresponding to a mean de Broglie wavelength of $\lambda = 0.56$ Å, or to $T = 83$ K, corresponding to $\lambda = 1.03$ Å. The atoms traveled through two slits, burned with a laser beam in a thin gold foil, and separated by 8 μm. The interference pattern was monitored by a detection system consisting of a secondary electron multiplier. The interference pattern of the detected atoms is shown in Fig. 6.1.

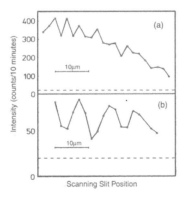

Figure 6.1 Interference pattern observed in the Carnal and Mlynek experiment with two different de Broglie wavelengths (a) $\lambda = 0.56$ Å and (b) $\lambda = 1.03$ Å. The dashed line represents the detector background with the atomic beam blocked in front of the entrance slit.

In summary of this section, we may conclude that for matter as well as for light, we must face up to the existence of a dual character: Matter behaves in some circumstances like a particle and in others like a wave.

What does it mean by "some circumstances"? One can say that this is a rather confusing statement. It means that the two models (particle and wave) complement each other.

6.2 Matter Wave Interpretation of Bohr's Model

De Broglie's wave–particle theory offered a much more satisfactory interpretation of the Bohr atom: Bohr's condition for angular momentum of the electron in a hydrogen atom is equivalent to a standing wave condition. The quantization of angular momentum $L = n\hbar$ means that

$$mvr = n\hbar$$

or

$$mv = \frac{nh}{2\pi r} . \tag{6.6}$$

However, if we employ de Broglie's postulate that the electron behaves as a wave, not as a particle

$$p = mv = \frac{h}{\lambda} , \tag{6.7}$$

and combine Eqs. (6.6) and (6.7), we find

$$n\lambda = 2\pi r , \qquad n = 1, 2, 3, \ldots \tag{6.8}$$

Thus, if one tries to represent the length in terms of wavelengths, the length of Bohr's allowed orbits $(2\pi r)$ exactly equals to an integer multiple of the electron wavelength $(n\lambda)$, as illustrated in Fig. 6.2. Hence, Bohr's quantum condition is equivalent to saying that an integer number of electron waves must fit into the circumference of a circular orbit.

The de Broglie wavelength of an electron in the smallest orbit turns out to be exactly equal to the circumference of the orbit predicted by Bohr. Similarly, the second and third orbits are found

(a)   **(b)**

 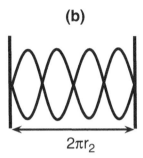

$2\pi r_1$ $2\pi r_2$

Figure 6.2   Example of standing waves in the length of the electrons' first $(n = 1)$ and second $(n = 2)$ orbit of lengths (a) $2\pi r_1$ and (b) $2\pi r_2$.

to contain two and three de Broglie wavelengths, respectively. From this picture, it now becomes clear why only certain orbits are allowed.

Note that de Broglie arrived to this conclusion from the fundamental matter–wave postulate, whereas Bohr assumed this property.

We can summarize that according to quantum wave mechanics:

(1) The electron motion in an atom is represented by standing waves.
(2) Since only certain wavelengths can now exist, the electron's energy can take on only certain discrete values.

Tutorial Problems

Problem 6.1 If, as de Broglie says, a wavelength can be associated with *every* moving particle, then why are we not forcibly made aware of this property in our everyday experience? In answering, calculate the de Broglie wavelength of each of the following "particles":

(a) A car of mass 2000 kg traveling at a speed of 120 km/h.
(b) A marble of mass 10 g moving with a speed of 10 cm/s.
(c) A smoke particle of diameter 10^{-5} cm (and a density of, say, 2 g/cm^3) being jostled about by air molecules at room temperature ($27°C = 300$ K). Assume that the particle has the

same translational kinetic energy as the thermal average of the air molecules

$$\frac{p^2}{2m} = \frac{3}{2} k_B T ,$$

where, as usual, k_B is Boltzmann's constant and T is the absolute temperature.

Problem 6.2 Show that the de Broglie wavelength of the electron in the ground state, $n = 1$, is equal to $2\pi a_o$, where a_o is the Bohr radius.

6.3 Definition of Wave Function

The idea that the electron's orbits in atoms correspond to standing matter waves was taken by Erwin Schrödinger in 1926 to formulate *quantum wave mechanics*. He introduced the idea of wave function $\Psi(\vec{r}, t)$ as the basic quantity in quantum wave mechanics. According to this idea, the wave function measures the wave disturbance of matter waves associated with a given system at time t and at a point $\vec{r}$.

But what exactly is this "wave function" and what is its physical meaning?

Before we explain the physical meaning of the wave function, consider a simple but important example in which we will determine what types of wave functions represent a physical system, and what are the physical consequences of choosing this particular form.

Suppose we have a free particle of mass m confined between two walls separated by a distance a. The motion of the particle along the x-axis may be represented by a harmonic wave

$$\Psi(x, t) = \Psi_{\max} \sin(kx) \sin(\omega t + \phi) , \qquad (6.9)$$

where $\omega = 2\pi \nu$ and $k = 2\pi / \lambda$.

However, only standing waves will exist for all times, and other waves will destructively interfere and disappear after some time t. Therefore, only standing waves can represent the particle confined between the walls. The condition required for a standing wave is

$$a = n\frac{\lambda}{2} , \qquad n = 1, 2, 3, \ldots \qquad (6.10)$$

from which we find that

$$k = \frac{2\pi}{\lambda} = \frac{n\pi}{a} .$$ (6.11)

Hence, the wave function takes the form

$$\Psi_n(x, t) = \Psi_{max} \sin\left(\frac{n\pi}{a}x\right) \sin(\omega t + \phi) ,$$ (6.12)

where the subscript n has been introduced to indicate that the particle is determined by an infinite number of separate wave functions.

Physically, all that Eq. (6.12) says is that a particle in a box is represented by a set of separate standing waves of different amplitudes. It means that the motion of the particle is quantized and is represented by an infinite (discrete) number of wave functions determined by different n. However, this is not the whole story. As a consequence of the quantization of the amplitude, the linear momentum is also quantized. Since

$$\lambda = \frac{2a}{n} ,$$ (6.13)

we can replace λ by h/p and obtain

$$p = n\frac{h}{2a} .$$ (6.14)

The momentum is related to the energy E, which gives

$$E = \frac{1}{2}\frac{p^2}{m} = n^2\frac{h^2}{8ma^2} \equiv E_n .$$ (6.15)

This indicates that the energy of the particle is also quantized. Thus, a particle confined between two walls cannot have any energy.

In summary, the main consequence of quantum confinement is that the particle is represented by a set of separate wave functions. In addition, the continuous energy spectrum of the particle in free space (unlimited space corresponding to $a \to \infty$) is radically modified and replaced by a *discrete* energy spectrum. The energy levels depend on the distance a and can be modified by appropriate choice of the value a.

6.4 Physical Meaning of Wave Function

The aforementioned theoretical analysis and diffraction experiments with particles definitely convinced us that a particle can *behave* like a wave. Remember, the particle itself is not a wave, but it behaves like a wave. If you say that the particle is a wave, then one may object to this interpretation and may say that it is problematic. Namely, a particle has mass and some of them have electric charge. Does this mean that the mass and charge of an electron, for example, are spread out over the extent of the wave? This would be crazy. It would mean that if we isolate just a part of the wave, we would obtain a fraction of an electron charge.

How then should we interpret an electron wave?

The answer is that the wave itself does not have any substance. It is a **probability wave**. When we talk about a particle wave, the amplitude of the wave at a particular point tells us the probability of finding the particle at that point.

We may recognize the close analogy between the wave function and, for example, the electric field amplitude. The wave function of a particle describes the probability distribution of a particle in space, just as the wave function of an electromagnetic field describes the distribution of the electromagnetic field in space.

In fact, the idea of the probabilistic interpretation of the physical phenomena, introduced by Born, Heisenberg, and Schrödinger, came from the probabilistic nature of the interference and diffraction effects. Simply, it is observed in diffraction experiments that the particles randomly distribute to form the diffraction pattern on the screen.

In addition, we know from the interference and diffraction involving an electromagnetic field that the intensity of the interference fringes is proportional to the square of the field amplitude, or alternatively to the probability that the waves interfere positively or negatively at some points.

In analogy to this theory of interference, Max Born suggested[a] that the quantity $|\Psi(\vec{r}, t)|^2 = \Psi^*(\vec{r}, t)\Psi(\vec{r}, t)$ is a measure of the

---

[a] Born was granted the Nobel Prize in 1954 for his fundamental research in quantum mechanics, especially for his statistical interpretation of the wave function.

probability density that the particle will be found at time t near the point $\vec{r}$. More precisely, the quantity $|\Psi(\vec{r}, t)|^2 dV$ is the **probability** that the particle will be found within a volume dV around the point $\vec{r}$ at which $|\Psi(\vec{r}, t)|^2$ is evaluated.[a]

Since $|\Psi(\vec{r}, t)|^2 dV$ is interpreted as the probability, it is normalized to one as

$$\int_V |\Psi(\vec{r}, t)|^2 dV = 1. \qquad (6.16)$$

The probabilistic formulation of quantum physics makes it completely different from classical physics. In classical physics, everything appears to have a definite position, a definite momentum, and a definite time of occurrence. The trajectory of a particle and the future behavior may be predicted with absolute certainty using Newton's laws. In other words, in Newtonian mechanics everything is predictable.

Let us consider two examples that illustrate differences between the classical and quantum behavior of a particle.

Example 1: *Classical behavior of a particle*

Suppose there is a force $\vec{F}$, which might be associated with a potential energy of a particle, acting on the particle initially at $t = t_0$ located at a point $\vec{r}_0$. The particle is moving with an initial velocity $\vec{v}_0$.

Using Newton's second law

$$\vec{F} = \frac{d\vec{p}}{dt} = m\frac{d^2\vec{r}}{dt^2}, \qquad (6.17)$$

we find with absolute certainty the particle's locality $\vec{r}(t)$ and velocity $\vec{v}(t)$ at any future time.

Moreover, we may predict the trajectory of the particle's motion. Let

$$\vec{F} = m\vec{g}. \qquad (6.18)$$

Then

$$m\vec{g} = m\frac{d\vec{v}}{dt}, \qquad (6.19)$$

---

[a]The wave function Ψ is complex, but the product $\Psi\Psi^*$ is always real and a positive quantity. The function Ψ^* is the complex conjugate of Ψ and is obtained from Ψ by replacing the complex parameter i by $-i$ whenever it appears in the function.

from which we find

$$\vec{v}(t) = \vec{v}_0 + \vec{g}t, \tag{6.20}$$

and

$$\vec{r}(t) = \vec{r}_0 + \vec{v}_0 t + \frac{1}{2}\vec{g}t^2. \tag{6.21}$$

Thus, if we know $\vec{r}_0$ and $\vec{v}_0$ at time $t = t_0$, we may predict the precise position $\vec{r}(t)$ and velocity $\vec{v}(t)$ of the particle at all future times $t > t_0$.

Example 2: *Quantum behavior of a particle*
Let us consider now the motion of the particle from the point of view of quantum physics, in which behavior of the particle is described by a wave function.

Consider the wave function at time $t = 0$ of a free particle confined between two walls (Eq. (6.12)). In this case, the probability density of finding the particle at a point x between the walls is given by

$$|\Psi(x, 0)|^2 = |\Psi_{max}|^2 \sin^2\left(\frac{n\pi}{a}x\right). \tag{6.22}$$

This formula shows that the probability of finding the particle at the point x is different from 1 (certain) and varies with the position x and the distance between the walls.

Thus, according to quantum physics, we are not able to predict with absolute certainty the initial $t = t_0$ position of the particle between the walls. Hence, we will not be able to predict with absolute certainty the trajectory of the particle, position $\vec{r}(t)$, and velocity $\vec{v}(t)$ at any future time $t > t_0$.

The dependence of the probability density on the position x between the walls for two different values of n is shown in Fig. 6.3. It is seen that for $n = 1$, the particle is more likely to be found near the center than the ends. For $n = 2$, the particle is most likely to be found at $x = a/4$, $x = 3a/4$, and the probability of finding the particle at the center is zero. The strong dependence of the probability on x is in contrast to the predictions of classical physics, where the particle has the same probability of being anywhere between the walls.

Such a behavior of the particle is another example of quantum effect without classical analog.

An aside: The readers may probably get a mixed feeling about these quantum ideas. It could be expected, as these quantum ideas

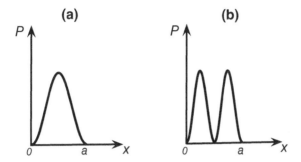

Figure 6.3 Probability density $P = |\Psi(x, 0)|^2$ as a function of the position x for a free particle confined between two walls, and (a) $n = 1$, (b) $n = 2$.

are not easy to grasp as they seem to contradict our intuitive understanding of the physical world.[a] It often leads people to question the physical models developed in physics. The probabilistic (statistical) nature of quantum physics is in itself a psychological barrier for many people. Even Einstein was inflexibly opposed to this statistical interpretation, which "leaves so much to chance." He never accepted these views and on some occasions dismissed them rather harshly:

> *I cannot believe that God plays dice with the cosmos.*
> —Albert Einstein

According to Einstein, quantum mechanics is an incomplete theory, fails to provide a complete description of physical reality, and that there exist certain *hidden variables* that, if known, would remove the necessity of the probabilistic interpretation of quantum mechanics.[b]

---

[a] If any reader knows why a particle confined between two walls behaves in such an unusual way, please write to us. We would like to know it, too.

[b] This problem is known as the Einstein–Podolsky–Rosen (EPR) paradox and is the most famous and powerful attack on the quantum wave mechanics. The detailed discussion of the paradox requires the knowledge of the advance quantum physics. However, the readers interested in learning more about the theory and experiments on the EPR paradox is referred to a recent review article by A. Zeilinger, Experiments in quantum mechanics, *Review Modern Physics*, **71**, S288 (1999).

Remember that the wave function of a particle does not tell us what the particle is, how it looks like, and what is its structure. The wave function $\Psi(\vec{r}, t)$ tells us about properties of the particle and how the particle *behaves* in space and time. Moreover, the wave function $\Psi(\vec{r}, t)$ is a mathematical construct only. It is a **probability wave**, which does not have any physical meaning.

Only $|\Psi(\vec{r}, t)|^2 = \Psi^*(\vec{r}, t)\Psi(\vec{r}, t)$ has physical meaning— **probability density**—and $|\Psi(\vec{r}, t)|^2 dV$ is the **probability** of finding the particle at time t in the volume dV around the point $\vec{r}$.

An interesting comment of a former student

"If only $|\Psi(\vec{r}, t)|^2 dV$ has physical meaning, why don't we use this probability rather than $\Psi(\vec{r}, t)$? Wouldn't it be simpler in quantum mechanics to use only the probability and forget about the wave function?"

We shall see in the next chapter that the answer to this question lies in the interference experiments, which show that the wave function itself plays an important role in the interpretation of many phenomena in quantum physics. Briefly, $\Psi(\vec{r}, t)$ describes the wave nature of a particle, and $|\Psi(\vec{r}, t)|^2 dV$ describes where we might find the particle.

Tutorial Problems

Problem 6.3 Determine where a particle is most likely to be found whose wave function is given by

$$\Psi(x) = \frac{1 + ix}{1 + ix^2}.$$

Problem 6.4 The wave function of a free particle at $t = 0$ is given by

$$\Psi(x, 0) = \begin{cases} 0 & x < -b, \\ A & -b \leq x \leq 3b, \\ 0 & x > 3b. \end{cases}$$

(a) Using the fact that the probability is normalized to one, i.e.,

$$\int_{-\infty}^{+\infty} |\Psi(x, 0)|^2 dx = 1,$$

find the constant A. (You can assume that A is real.)

(b) What is the probability of finding the particle within the interval $x \in [0, b]$ at time $t = 0$?

Problem 6.5 The state of a free particle at $t = 0$ confined between two walls separated by a is described by the following wave function:

$$\Psi(x, 0) = \Psi_{max} \sin\left(\frac{n\pi}{a}x\right), \qquad 0 \le x \le a,$$

$$\Psi(x, 0) = 0, \qquad x > a, \quad \text{and} \quad x < 0.$$

(a) Find the amplitude Ψ_{max} using the normalization condition.

(b) What is the probability density of finding the particle at $x = 0$, $a/2$, and a? How does the result depend on n?

(c) Calculate the probability of finding the particle in the regions $\frac{a}{2} \le x \le a$ and $\frac{3a}{4} \le x \le a$, for $n = 1$ and $n = 2$.

Problem 6.6 The time-independent wave function of a particle is given as

$$\Psi(x) = Ae^{-|x|/\sigma},$$

where A and σ are constants.

(a) Sketch this function and find A in terms of σ such that $\Psi(x)$ is normalized.

(b) Find the probability that the particle will be found in the region $-\sigma \le x \le \sigma$.

6.5 Phase and Group Velocities of Matter Waves

We have already learned that light has a dual character, and after de Broglie and the interference experiments with particles, we believe that matter also has dual character. A particular consequence of this is that a particle confined between two walls and treated as a matter wave cannot have any energy. We remember from a previous chapter that the radiation in a box exhibits a similar property, which led Planck to introduce the concept of quantization of energy. In this chapter, we continue our study of the duality problem and turn our attention to the following question: What is the velocity with which the probability moves in space and time? To answer this question, we shall introduce the concept of phase and group velocities and will examine the variation of the velocity of matter waves with frequency of these waves. In the wave mechanics, this effect is called the *dispersion*.

We have seen that the radiation and matter contain within their structures both wave and particle concepts.

Waves concept

$$\lambda = \frac{h}{p} = \frac{h}{mv} , \tag{6.23}$$

where v is the velocity of a particle of the mass m.

Particles concept

$$E = mc^2 = h\nu . \tag{6.24}$$

Hence, if we accept that there is an analogy between the matter waves and radiation, we shall find that the velocity of the matter waves is

$$u = \lambda\nu = \frac{h}{mv} \frac{mc^2}{h} = \frac{c^2}{v} . \tag{6.25}$$

Since, $v < c$, we see that the velocity of the matter waves is always greater than the speed of light in vacuum, i.e., $u > c$. Thus, u and v are never equal for a moving particle.

This result seems disturbing because it appears that the matter waves, if treated as an analog of radiation, would propagate faster than the speed of light in vacuum and would not be able to keep up with particles whose motion they govern.

However, the velocity u is a *phase velocity* of the matter waves, which is the velocity of the wave front, not its amplitude. The maximum of the amplitude of a given wave can propagate at different velocities, called *group velocity*. At this velocity, the energy (information) is transmitted. Usually, $v_g = u$, but in the case of dispersion, $u(\nu)$, the group velocity $v_g < u$. Thus, the matter wave should be dispersive to match the requirement of $v_g < u$ when $u > c$.

Thus, an obvious question arises from the foregoing: Are the matter waves dispersive?

Let us answer this by first defining the phase and group velocities.

In order to introduce the concept of group velocity and to show that matter waves are dispersive, we have to assume that a matter wave is composed at least of two harmonic waves of slightly different k and ω. We shall explain it later in the book why we have to assume that a matter wave is composed of a group of harmonic waves.

Suppose we have two harmonic waves of slightly different k and ω and propagating in the same direction. Let

$$k_1 = k_0 + \Delta k, \qquad \omega_1 = \omega_0 + \Delta \omega,$$
$$k_2 = k_0 - \Delta k, \qquad \omega_2 = \omega_0 - \Delta \omega, \qquad (6.26)$$

The resulting wave is obtained by taking a linear superposition of the two waves

$$\Psi(\vec{r}, t) = \frac{1}{2} e^{i(\vec{k}_1 \cdot \vec{r} - \omega_1 t)} + \frac{1}{2} e^{i(\vec{k}_2 \cdot \vec{r} - \omega_2 t)}. \qquad (6.27)$$

Then, using Eq. (6.26) and Euler's formula $(e^{\pm ix} = \cos x \pm i \sin x)$, we obtain

$$\Psi(\vec{r}, t) = \frac{1}{2} e^{i[(k_0 + \Delta k)\hat{\kappa} \cdot \vec{r} - (\omega_0 + \Delta \omega)t]} + \frac{1}{2} e^{i[(k_0 - \Delta k)\hat{\kappa} \cdot \vec{r} - (\omega_0 - \Delta \omega)t]}$$
$$= e^{i(k_0 \hat{\kappa} \cdot \vec{r} - \omega_0 t)} \cos (\Delta k \hat{\kappa} \cdot \vec{r} - \Delta \omega t), \qquad (6.28)$$

where $\hat{\kappa} \cdot \vec{r}$ is the distance the wave propagated, and $\hat{\kappa}$ is the unit vector in the $\vec{k}$ direction.

We see from Eq. (6.28) that in time t, the fast varying function propagates a distance

$$\hat{\kappa} \cdot \vec{r} = \frac{\omega_0}{k_0} t = ut, \qquad (6.29)$$

whereas the envelope propagates a distance

$$\hat{\kappa} \cdot \vec{r} = \frac{\Delta \omega}{\Delta k} t = \frac{d\omega}{dk} t = v_g t. \qquad (6.30)$$

Hence, the envelope propagates at velocity $v_g = d\omega/dk$, which is called the **group velocity**.

We see from Eq. (6.28) that the probability density $|\Psi(\vec{r}, t)|^2$ is independent of the phase velocity; it depends only on the group velocity

$$|\Psi(\vec{r}, t)|^2 = \cos^2 \Delta k \left(\vec{\kappa} \cdot \vec{r} - v_g t \right) . \tag{6.31}$$

Thus, the probability moves with the group velocity.

The envelope forms a so-called **wave packet** and, as we have already seen, the amplitude of the wave packet propagates with velocity v_g.

We are now at the position to answer the main question of whether matter waves are dispersive or not. To check this, consider the energy of a particle as

$$E = \frac{1}{2m}p^2 = \frac{\hbar^2}{2m}k^2 . \tag{6.32}$$

If the energy of the particle is quantized, $E = \hbar\omega$, and then

$$\hbar d\omega = \frac{\hbar^2}{2m}2kdk , \tag{6.33}$$

from which we find that

$$v_g = \frac{d\omega}{dk} = \frac{\hbar k}{m} \neq u . \tag{6.34}$$

Hence, if $E = \hbar\omega$, then the matter waves are dispersive.

When $v_g < u$, we say that the matter waves exhibit *normal dispersion*, whereas $v_g > u$ is regarded as *anomalous dispersion*. When $v_g = u$, there is no dispersion.

Let us consider two examples that illustrate some of the concepts just introduced.

Worked Example

What is the group velocity of the wave packet associated with a particle moving with velocity v?

Solution

From the definition of group velocity, we have

$$v_g = \frac{d\omega}{dk} = 2\pi \frac{dv}{dk} = 2\pi \frac{d(hv)}{d(hk)} = \frac{dE}{dp} , \tag{6.35}$$

where $p = \hbar k$. However,

$$E^2 = m_0^2 c^4 + p^2 c^2 . \tag{6.36}$$

Thus,

$$2E\,dE = 2pc^2\,dp . \tag{6.37}$$

from which we find that

$$\frac{dE}{dp} = \frac{pc^2}{E} = \frac{mvc^2}{mc^2} = v . \tag{6.38}$$

Hence, $v_g = v$, the group velocity is equal to the velocity of the particle. In other words, the velocity of the particle is equal to the group velocity of the corresponding wave packet.

Worked Example

The dispersion relation for free relativistic electron waves is

$$\omega_k = \sqrt{c^2 k^2 + (mc^2/\hbar)^2} . \tag{6.39}$$

(a) Calculate expressions for the phase velocity u and group velocity v_g of these waves and show that their product is constant, independent of k.
(b) From the result (a), what can you conclude about v_g if $u > c$?

Solution

(a) From the definition of the phase velocity, we find

$$u = \frac{\omega_k}{k} = \sqrt{c^2 + \left(\frac{mc^2}{k\hbar}\right)^2} . \tag{6.40}$$

We see that the phase velocity $u > c$.
From the definition of the group velocity, we find

$$v_g = \frac{d\omega_k}{dk} = \frac{1}{2}\frac{2c^2 k}{\sqrt{c^2 k^2 + (mc^2/\hbar)^2}}$$

$$= \frac{c^2 k}{ck\sqrt{1 + (mc/k\hbar)^2}} = \frac{c}{\sqrt{1 + (mc/k\hbar)^2}} . \tag{6.41}$$

Thus, the group velocity is less than c ($v_g < c$) as it must be since, according to the theory of relativity, energy or a signal cannot be propagated with a velocity exceeding c.

In addition, the product

$$uv_g = \frac{c^2 k}{\omega_k} \frac{\omega_k}{k} = c^2 \tag{6.42}$$

is constant and independent of k. Thus, relativity is still all right.

(b) We see from (a) that in general for dispersive waves for which $u > c$, the group velocity $v_g < c$. Only when $u = c$, the group velocity

$$v_g = c.$$

To summarize, we have learned that

(1) Matter waves are dispersive.
(2) The matter wave associated with a particle is in the form of a wave packet.
(3) The amplitude of a matter wave (wave packet) moves with the group velocity.
(4) The group velocity of the matter wave associated with a moving particle travels with the same velocity as the particle.

In the next step of our efforts to understand the fundamentals of quantum physics, we will explain why in quantum physics a localized particle is represented by a superposition of wave functions (wave packet) rather than a single harmonic wave function. Important steps on the way to understand the concept of wave packets are the uncertainty principle between the position and momentum of the particle, and the superposition principle.

Tutorial Problems

Problem 6.7 In one of the chapters, we calculated the phase velocity u using the relativistic formula for the energy. Calculate the phase velocity for the non-relativistic case. Does the relativistic result for u tends to the corresponding non-relativistic result as the velocity of the particle becomes small compared to the speed of light?

Problem 6.8 We know that the group velocity v_g of the wave packet of a particle of mass m is equal to the velocity v of the particle. Show that the total energy of the particle is $E = \hbar\omega$, the same which holds for photons.

Problem 6.9 If the group velocity of a wave packet is given by $v_g = 3u$, where u is the phase velocity, how does u depend on frequency ω?

> *What the electron is doing during its journey in the interferometer? During this time the electron is a great smoky dragon, which is only sharp at its tail (at the source) and at its mouth, where it bites the detector.*
>
> —J. A. Wheeler

6.6 Heisenberg Uncertainty Principle

In quantum physics, we usually work with the wave function Ψ, whose $|\Psi|^2 dV$ describes a probability that a given object, e.g., a particle confined to a volume dV, is moving with a velocity v_0. A probability different from 1 (certain) means that we are not precisely sure that the velocity of the particle is v_0. We may say that the velocity is v_0 with some error Δv_0, which is called standard deviation or variance or simply **uncertainty**.

The same argument applies to a measurement of position of the particle. In fact, it applies to the measurement of any quantity in physics. Measurements of some quantities are independent of each other, but some are related (correlated), i.e., a measurement of one of the quantities affects the measurement of the other. We call this relation the **Heisenberg uncertainty principle**.

The uncertainty principle can be derived in a variety of ways. It applies in both classical and quantum physics. Let us obtain it by considering a typical diffraction experiment shown in Fig. 6.4.

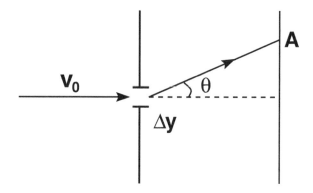

Figure 6.4 Schematic diagram of a diffraction experiment. A beam of particles emerging from the slit of width Δy interferes to form an interference pattern on the observation screen.

The position of the first ($n = 1$) minimum in the diffraction pattern is given by the diffraction formula

$$\sin\theta = \frac{\lambda}{\Delta y}. \tag{6.43}$$

In order to reach the point A, the particle initially moving in the x-direction has to gain a velocity in the y-direction, such that

$$\sin\theta = \frac{\Delta v_y}{v_0}, \tag{6.44}$$

where Δv_y is a change (gain) of the velocity of the particle in the y-direction.

By comparing Eqs. (6.43) and (6.44), we find that

$$\frac{\Delta v_y}{v_0} = \frac{\lambda}{\Delta y}, \tag{6.45}$$

which we may write as

$$\Delta v_y \Delta y = v_0 \lambda. \tag{6.46}$$

This relation is often called the *classical uncertainty relation*. The uncertainty is limited by the wavelength λ and can be changed by changing λ.

When we apply the de Broglie postulate

$$\lambda = \frac{h}{p} = \frac{h}{mv_0}, \tag{6.47}$$

we find that the relation (6.46) takes the form

$$\Delta p_y \Delta y = h, \qquad\qquad (6.48)$$

where $\Delta p_y = m\Delta v_y$ is the uncertainty in the momentum of the particle.[a]

This is the *quantum uncertainty relation*. The precision of measurement of a given quantity is independent of any experimental parameter. It is limited by the Planck constant h.

The relation (6.48) is one form of the **uncertainty principle** first obtained by Werner Heisenberg[b] in 1927. It states that it is impossible to measure the momentum p_y and position y of a particle simultaneously with the same precision. If we know more about one of the two quantities, the less we know about the other. For example, if the particle is completely unlocalized, $\Delta y \to \infty$, the momentum is certain, $\Delta p_y \to 0$, and vice versa; if the momentum is completely random, $\Delta p_y \to \infty$, the position is certain, $\Delta y \to 0$. Two physical (measurable) quantities related through the uncertainty principle are called *complementary observables*.

An aside: The Heisenberg uncertainty principle has created a long debate on the validity of quantum physics. Most scientists interpret physical phenomena as events taking place "out there," independent of any measurement or observation. At the same time, quantum theory stands in conflict with such naive notions of reality. As we have already learned, the Heisenberg uncertainty principle sets a limit on the precision with which two complementary observables can be measured. For example, a measurement of momentum of a particle disturbs the position of the particle. To many people, this is an unsatisfactory feature of quantum physics. The most notable objector, of course, was Einstein, whose concern about the uncertainty principle is expressed in his famous statement:

---

[a]The arguments used in the derivation of Eq. (6.48) are somewhat rough. A more concrete treatment, which will be presented in Section 10.6, results in the following inequality, called the Heisenberg inequality

$$\Delta p_y \Delta y \geq h/4\pi .$$

Note that the relation (6.48) satisfies the Heisenberg inequality as $h > h/4\pi$.

[b]Heisenberg was granted the Nobel Prize in 1932 for the creation of quantum mechanics.

Is the state of the Universe disturbed if a mouse looks at it?.

In summary, remember that the Heisenberg uncertainty principle is not a statement about the inaccuracy of measurement instruments, nor a reflection on the quality of experimental methods. It arises from the wave properties inherent in the quantum mechanical description of nature. Even with perfect instruments and techniques, the uncertainty is inherent in the nature of things and quantifies the inability to precisely locate them.

Discussion Problem

Problem D1 Regarding uncertainty principle, a student asked a question: "If I do not move, does it mean that I am everywhere"? How would you answer this question?

Tutorial Problems

Problem 6.10 Monochromatic light, such as produced by lasers, is used to determine the position of small objects such as particles or single trapped atoms. Suppose that visible light of wavelength $\lambda = 5 \times 10^{-7}$ m is used to determine the position of an electron within the wavelength of the light. What is the uncertainty in the electron's velocity?

Problem 6.11 *Energy and time uncertainty relation*
The time required for a wave packet to move the distance equal to the width of the wave packet is $\Delta t = \Delta x / v_g$, where Δx is the width of the wave packet. Show that the time Δt and the uncertainty in the energy of the particle satisfy the uncertainty relation

$$\Delta E \, \Delta t = h \,,$$

where $\Delta E = \hbar \Delta \omega$.

*I cannot define the real problem therefore I suspect there is
no real problem, but I am not sure there is no real problem.*
—R. P. Feynman

6.7 Superposition Principle

According to quantum physics, we may associate with each particle
a wave function to determine its position $\vec{r}$ at a given time t:

$$\Psi(\vec{r}, t) = A e^{i(\vec{k}\cdot\vec{r} - \omega t)}. \tag{6.49}$$

Since the probability density $|\Psi(\vec{r}, t)|^2 = |A|^2 = $ const., i.e., it
is independent of $\vec{r}$ and t, we see that the particle is completely
unlocalized in space and can be found anywhere in space with
the same probability. However, we know from everyday life that
particles are localized in space and their position can be given with
some approximation. In other words, particles are partly localized.
Therefore, a wave function such as (6.49) cannot represent a real
physical system.

How to get out of this dilemma?

We may resolve this dilemma by referring to the uncertainty
principle. According to the uncertainty principle, a particle partly
localized in space ($\Delta\vec{r} < \infty$) has an uncertainty in momentum
($\Delta\vec{p} \neq 0$). Hence, if

$$\Psi_1(\vec{r}, t) = A_1 e^{i(\vec{p}_1\cdot\vec{r}/\hbar - \omega_1 t)} \tag{6.50}$$

is a wave function of the particle located at $\vec{r}$, then

$$\Psi_2(\vec{r}, t) = A_2 e^{i(\vec{p}_2\cdot\vec{r}/\hbar - \omega_2 t)}, \tag{6.51}$$

where $|\vec{p}_1 - \vec{p}_2| \leq \Delta\vec{p}$ is also a wave function of the particle.

Moreover, any linear combination (superposition) of the two
wave functions is also a wave function of the particle,[a] i.e.,

$$\Psi(\vec{r}, t) = a\Psi_1(\vec{r}, t) + b\Psi_2(\vec{r}, t), \tag{6.52}$$

where a and b are complex numbers.

---

[a]We will see in Chapter 7 that this conclusion arises from the linearity in Ψ of
the Schrödinger equation, which determines the wave function of a given particle.
Namely, the linearity simply means that if Ψ_1 and Ψ_2 are solutions to the Schrödinger
equation, then an arbitrary linear combination of Ψ_1 and Ψ_2 is also a solution.

Equation (6.52) is an example of the *superposition principle*, which, in general, holds for an arbitrary number of wave functions.

Thus, a single wave function cannot represent a particle of a given momentum. Instead, we expect that the particle is represented by a superposition of wave functions. Hence, we may conclude that the wave function of a particle is represented by the sum of sinusoidal waves $\exp[i(\vec{k} \cdot \vec{r} - \omega_k t)]$. For a continuous set of wave functions, the sum is of course an integral

$$\Psi(\vec{r}, t) = \int_k A(\vec{k}) e^{i(\vec{k} \cdot \vec{r} - \omega_k t)} d^3 k, \tag{6.53}$$

where $d^3 k$ is the element of volume in $\vec{k}$-space (momentum space). In other words, the set contains an infinite number of waves with continuously varying wave number k.

One can see from Eq. (6.53) that the mathematics used in carrying out the procedure of obtaining a superposition wave function involves the Fourier transformation (Fourier integral). If the superposition function is known, the amplitude $A(\vec{k})$ can be found employing the inverse Fourier transformation:

$$A(\vec{k}) = \frac{1}{\sqrt{2\pi}} \int_V \Psi(\vec{r}, t) e^{-i(\vec{k} \cdot \vec{r} - \omega_k t)} d^3 \vec{r}. \tag{6.54}$$

In summary, the superposition principle is at the heart of quantum mechanics and is referred by many physicists as really "the only mystery" of quantum mechanics. It is in complete contrast with classical mechanics, where a superposition of two states would be a complete nonsense as it would imply that a particle could simultaneously occupy two or more points in space. According to quantum physics, a particle can exist in two or more states at the same time. Furthermore, if more particles are involved, we encounter the idea of a multi-particle superposition or non-separability, called *quantum entanglement*. Clearly, the superposition principle confronts us with some basic questions of interpretation of physical reality. The superposition principle and entanglement have been exploited in recent years in three important applications. The first is *quantum cryptography*, where a communication signal between two people can be made completely secure from eavesdroppers. The second is *quantum communication*, where capacities of transmission lines can be increased in comparision to that of classical transmission systems. The third is a proposed device called *quantum computer,* where all possible calculations could be carried out simultaneously.

6.8 Quantum Interference

A beautiful illustration of the superposition principle is the interference phenomenon. Although interference is a classical wave phenomenon, it is also at the heart of quantum mechanics. It is usually illustrated with Young's double-slit experiment, in which a beam of light is divided at two narrow slits into two beams that, after traveling separately for some distance, are recombined at an observation point.

If there is a small path difference between the beams, interference fringes are observed at the observation (recombination) point. A schematic diagram of Young's double-slit experiment is shown in Fig. 6.5.

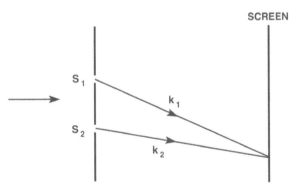

Figure 6.5 Schematic diagram of Young's double-slit experiment. Two light beams emerging from the slits S_1 and S_2 are brought together at the observing screen and the resulting light intensity is measured at various positions.

The observed intensity is a pattern of alternating bright and dark strips, called *interference fringes.* Understanding the reason for this interference is not difficult. It may be explained using classical description of the process. The paths from each slit to a given point on the observation screen are not necessarily equal, so the beams traveling from the slits arrive with different phases of propagation. Due to the phase difference, the two beams will yield constructive interference (bright strips) if they are in phase at the observation point and destructive interference (dark strips) if they are 180° out of phase with each other.[a]

The interpretation of the interference phenomenon becomes more subtle when the light beams are replaced by individual photons or particles. What happens is that when single photons or particles are used and the experiment is repeated many times, it is observed that the resulting pattern on the screen is identical to that observed with the light beams. This seems to imply that the individual photons had passed through both slits at the same time and interfered with themselves. Dirac first noticed this strange behavior, and his statement is now cited as the famous phrase "each photon interferes only with itself. Interference between different photons never occurs."

How could the photon travel simultaneously through both slits?

This confusion is explained in terms of the superposition principle resulting from the lack of information about the slit through which the photon has been transferred to the screen. The interference pattern observed in Young's double-slit experiment results from a superposition of the probability amplitudes for the photon to take either of the two possible pathways. After the interaction of the photon with the slits, the system of the two slits and a photon is a *single* quantum system. In this case, the wave function of the photon detected at the screen is a sum of two wave functions corresponding to the two possible pathways the photon traveled to the screen

$$\Psi(\vec{R}) = \Psi_1(\vec{R}) + \Psi_2(\vec{R}), \tag{6.55}$$

where $\vec{R}$ is the position of the detector on the screen.

The probability of detecting the photon at R is obtained by calculating the square of the absolute value of $\Psi(\vec{R})$:

$$\left|\Psi(\vec{R})\right|^2 = \left|\Psi_1(\vec{R}) + \Psi_2(\vec{R})\right|^2 = \left|\Psi_1(\vec{R})\right|^2 + \left|\Psi_2(\vec{R})\right|^2$$
$$+ 2\text{Re}\left[\Psi_1^*(\vec{R})\Psi_2(\vec{R})\right], \tag{6.56}$$

---

[a] Certain fundamental conditions must be satisfied to obtain interference fringes: the beams must have the same intensity, frequency, wavelength, and polarization.

where "Re" denotes the real part. In this equation, the term $|\Psi_1(\vec{R})|^2$ is the probability for the photon to pass through slit 1; the term $|\Psi_2(\vec{R})|^2$ is the probability for the photon to pass through slit 2; and the last term is the interference (superposition) between these two probabilities. Since the wave functions depend on the position of the slits, $\Psi_i(\vec{R}) \sim e^{i\vec{k}_i \cdot \vec{R}}$, we see that $|\Psi(\vec{R})|^2$ is a cosine function of position $\vec{R}$ and the recorded signal will show a spatial modulation (interference fringes) on the screen.

Note that only the last term in Eq. (6.56) varies with the position on the screen. Thus, this term is the only one responsible for the interference fringes observed on the screen. It shows that interference is a clear example of non-separability or entanglement in quantum mechanics. Of course, the concept of a photon as a discrete localized object is not helpful in understanding this experiment.

When we close one of the slits, say slit 1, $\Psi_1(\vec{R}) = 0$, and then the probability (6.56) reduces to

$$\left|\Psi(\vec{R})\right|^2 = \left|\Psi_2(\vec{R})\right|^2 , \qquad (6.57)$$

which shows that in this case no interference fringes are observed. When one of the slits is closed, we definitely know through which slit the photon traveled to the screen. This example is a clear illustration that the observation of interference fringes and the acquisition of which way the photon (information) was transferred are mutually exclusive. This problem is often referred to by the German phrase "welcher weg" (which way). Thus, interference is always a manifestation of the intrinsic indistinguishability of two possible paths of the detected photon. This indistinguishability is an example of Bohr's principle of *complementarity* that interference and which way information are mutually exclusive concepts.[a]

An aside: The welcher weg problem has created many discussions on the validity of the principle of complementarity. Einstein proposed modifying Young's double-slit experiment by using freely moving slits. A light beam, or a particle, arriving at a point on the screen must have changed momentum when passing through the

---

[a]Readers wishing to learn more about quantum interference, phenomena involving quantum interference, and some interference experiments are referred to a book by Z. Ficek and S. Swain, *Quantum Interference and Coherence: Theory and Experiments* (Springer, Berlin, New York, Heidelberg, 2005).

slits. Since the paths of the light beams traveling from the slits to the point on the screen are different, the change of the momentum at each slit must be different. Einstein's proposal was simply to observe the motion of the slits after the light beam traversed them. Depending on how rapidly they were moving, one could deduce through which slit the light beam had passed and, simultaneously, one could observe an interference pattern. If this were possible, it would be a direct contradiction of the principle of complementarity. However, Bohr proved that this proposal was deceptive in the sense that the position of the recoiling slits was subject to some uncertainty provided by the uncertainty principle. As a result, if the slits are moveable, a random phase is imparted to the light beams, and hence the interference pattern disappears.

Feynman in his proposal for a welcher weg experiment suggested replacing the slits in the usual Young's experiment by electrons. Because electrons are charged particles, they can interact with the incoming electromagnetic field. Feynman suggested putting a light source symmetrically between the slits. If the light beam is scattered by an electron, the direction of the scattered beam will precisely determine from which electron the beam has been scattered. In this experiment, the momentum of the electrons and their positions are both important parameters. To determine which electron had scattered the light beam and at the same moment observe interference, the momentum and the position of the electron would have to be measured to accuracies greater than allowed by the uncertainty principle.

6.9 Wave Packets

We have already learned that a free particle is not represented by a single wave function but rather by a superposition of closely related wave functions. Such a superposition is named a free particle **wave packet**. A typical wave packet of a free particle is given by Eq. (6.53). It also shows that the momentum and then also the position distribution are roughly pictured by the behavior of $|A(\vec{k})|^2$.

We now consider the motion of a free particle wave packet. For simplicity, we consider the motion in one dimension. In this case,

$$\Psi(\vec{r}, t) \rightarrow \Psi(x, t) = \int_{-\infty}^{\infty} A(k) e^{i(kx - \omega_k t)} dk, \qquad (6.58)$$

where $k = k_x$ and the frequency ω_k is different for different k.

Let us assume that $A(k)$ is appreciable only for k values that lie in an interval $k_0 - \frac{1}{2}\Delta k$, $k_0 + \frac{1}{2}\Delta k$, where $\Delta k/k_0 \ll 1$ and k_0 corresponds to the central (maximum) momentum.

Consider first the shape of the wave packet at $t = 0$. For $t = 0$, the wave packet (6.58) reduces to

$$\Psi(x, 0) = \int_{-\infty}^{\infty} A(k) e^{ikx} dk. \qquad (6.59)$$

We see that waves with different k have different phases.

Is there any relation between different phases and different positions of the fronts of the superimposed waves?

If Δx is the displacement of x from $x = 0$, we may calculate under what condition the particle can be found in the region Δx. Since the phases of the waves are different, calculate the maximal and minimal phases of the packet:

$$\Delta x \left(k_0 - \frac{1}{2}\Delta k \right), \qquad \text{minimal}$$

$$\Delta x \left(k_0 + \frac{1}{2}\Delta k \right), \qquad \text{maximal}.$$

Waves with different phases will interfere with each other. The maximum of interference appears for the difference between the phases equal to 2π. Thus,

$$\Delta x \Delta k = 2\pi. \qquad (6.60)$$

Hence, the particle can be found at points for which $\Delta x = 2\pi/\Delta k$, i.e., determined by the uncertainty relation.

Now we will check how the packet moves in time.

To do this, we may expand the frequency ω_k, appearing in Eq. (6.58), into a Taylor series about $\omega_{k_0} = \omega_0$ corresponding to the maximum of the momentum at $k = k_0$. By taking $k = k_0 + \beta$, where β is a small displacement from k_0, the frequency can be expanded as

$$\omega_k = \omega_{k_0 + \beta} = \omega_0 + \left(\frac{d\omega}{d\beta} \right)_{k_0} \beta + \frac{1}{2} \left(\frac{d^2\omega}{d\beta^2} \right)_{k_0} \beta^2 + \dots \quad (6.61)$$

Provided β is small enough, we take only first two terms of the series. Then substituting the expansion to $\Psi(x, t)$, we obtain

$$\Psi(\vec{r}, t) = e^{i(k_0 x - \omega_0 t)} \int_{-\infty}^{\infty} d\beta\, A\,(k_0 + \beta)\, e^{i\beta(x - v_g t)}, \qquad (6.62)$$

where $v_g = \left(\frac{d\omega}{d\beta}\right)_{k_0}$ is the group velocity of the packet.

If we increase x by Δx, i.e., $x \to x + \Delta x$, then

$$e^{i\beta(x - v_g t)} = e^{i\beta x} e^{i\beta(\Delta x - v_g t)}. \qquad (6.63)$$

Thus, for $\Delta x = v_g t$, we obtain the same packet as for $t = 0$, but shifted by $v_g t$. The conclusion then is that the group velocity is the velocity of the packet moving as a whole.

If we include the third term of the Taylor expansion (6.61), we get

$$\Psi(\vec{r}, t) = e^{i(k_0 x - \omega_0 t)} \int_{-\infty}^{\infty} d\beta\, A\,(k_0 + \beta)$$

$$\times \exp i\beta \left[x - \left(v_g + \left(\frac{dv_g}{d\beta}\right)_{k_0} \beta \right) t \right]. \qquad (6.64)$$

The term $v_g + \left(\frac{dv_g}{d\beta}\right)_{k_0} \beta$ plays the role of the velocity of the wave packet, which now depends on β. Thus, different parts of the wave packet will move with different velocities, leading to a spreading of the wave packet. This spreading is due to dispersion that v_g depends on β.

We can now summarize to give the connection between the group velocity and the phase velocity, and the role of dispersion.

$$\text{Phase velocity} \qquad u = \frac{\omega}{k},$$

$$\text{Group velocity} \qquad v_g = \frac{d\omega}{dk}.$$

Hence

$$v_g = \frac{d\omega}{dk} = \frac{d}{dk}(ku) = u + k\frac{du}{dk}. \qquad (6.65)$$

Thus, v_g depends on k when $\frac{du}{dk} \neq 0$, i.e., when the phase velocity depends on k. The dependence of v_g on k is called *dispersion*.

In addition, we can say that the spread of the wave packet is due to the dependence of the phase velocity on k [$v_g \neq u$ when $\frac{du}{dk} \neq 0$].

Let us summarize what we have learned in this lecture:

(1) In quantum physics, localized particles are represented by a superposition of wave functions (so-called wave packets) rather than a single harmonic wave function.
(2) The fact that the wave function of a particle must be represented by a group of waves of different momenta suggests that there is a fundamental limit to the accuracy with which the momentum and position of the particle can be measured.
(3) The maximum of a wave packet moves through space with the group velocity.
(4) The group velocity of the wave packet associated with a moving particle is the same as the velocity of the particle.
(5) Since the matter waves are dispersive, a wave packet spreads out as time progresses, which means that the position becomes more uncertain.

Revision Questions

Question 1 Define the phase and group velocities and what do they describe?

Question 2 Prove that matter waves are dispersive.

Tutorial Problems

Problem 6.12 The amplitude $A(k)$ of the wave function

$$\Psi(x, t) = \int_{-\infty}^{+\infty} A(k)e^{i(kx - \omega_k t)}dk$$

is given by

$$A(k) = \begin{cases} 1 & \text{for} \quad k_0 - \frac{1}{2}\Delta k \leq k \leq k_0 + \frac{1}{2}\Delta k, \\ 0 & \text{for} \quad k > k_0 + \frac{1}{2}\Delta k, \quad \text{and} \quad k < k_0 - \frac{1}{2}\Delta k. \end{cases}$$

(a) Show that the wave function can be written as

$$\Psi(x, t) = \frac{\sin z}{z} \Delta k \, e^{i(k_0 x - \omega_0 t)} ,$$

where $z = \frac{1}{2}\Delta k(x - v_g t)$.

(b) Sketch the function $f(z) = \sin z/z$ and find the width of the main maximum of $f(z)$.

(*Hint:* For $f(z)$, one might define a suitable width as the spacing between its first two zeros.)

Problem 6.13 Calculate $A(k)$ (inverse Fourier transform)

$$A(k) = \frac{1}{\sqrt{2\pi}} \int_{-\infty}^{+\infty} \Psi(x, 0) e^{-ikx} dx .$$

of the triangular wave packet

$$\Psi(x, 0) = \begin{cases} 1 + \frac{x}{b} & -b \le x \le 0 , \\ 1 - \frac{x}{b} & 0 < x < b , \\ 0 & \text{elsewhere} . \end{cases}$$

Draw qualitative graphs of $A(k)$ and $\Psi(x, 0)$. Next to each graph, write down its approximate "width."

Problem 6.14 The wave function of a particle is given by a wave packet

$$\Psi(x, t) = \int_{-\infty}^{+\infty} A(k) e^{i(kx - \omega_k t)} dk .$$

Assuming that the amplitude $A(k) = \exp(-\alpha|k|)$, show that the wave function is in the form of a Lorentzian

$$\Psi(x, t) = \frac{2\alpha}{\alpha^2 + (x - v_g t)^2} .$$

(*Hint:* Expand k and ω_k in a Taylor series around $k_0 = \omega_0 = 0$.)

Chapter 7

Non-Relativistic Schrödinger Equation

The major problem in quantum physics is to find the wave function of a given physical system and to understand (predict) how the wave function of the system evolves in time, or how it changes under external influences.

In 1926, Erwin Schrödinger predicted that the wave function of a given physical system might be completely determined if the total energy of the system was known.[a] He formulated an equation of motion for the wave function of a physical system, which is called the **Schrödinger equation**. It is the basic relationship for determining the evolution of the wave function and possible energies of a given physical system.

We shall try to find a differential equation for the wave function of a particle assuming that only the energy of the particle is known. Since the equation represents a real physical system, it must satisfy the following conditions:

- The equation must be linear.
- Coefficients appearing in this equation should only depend on the parameters characteristic of the particle.

---

[a]Schrödinger was granted the Nobel Prize in 1933 for his discovery of new productive forms of atomic theory.

Quantum Physics for Beginners
Zbigniew Ficek
Copyright © 2016 Pan Stanford Publishing Pte. Ltd.
ISBN 978-981-4669-38-2 (Hardcover), 978-981-4669-39-9 (eBook)
www.panstanford.com

We will limit our considerations to the non-relativistic case only.

7.1 Schrödinger Equation of a Free Particle

First, we will consider the case of a free particle moving in one dimension. The wave function of a free particle moving along, say the x-axis, is given by

$$\Psi(x, t) = \Psi_{max} e^{i(kx - \omega t)} . \tag{7.1}$$

On the other hand, the energy and momentum of a free particle are related by

$$E = \frac{1}{2m} p_x^2 . \tag{7.2}$$

Since $E = \hbar\omega$ and $p_x = \hbar k$, we can express the energy in terms of the wave parameters

$$\omega = \frac{\hbar}{2m} k^2 . \tag{7.3}$$

Note that

(1) Taking the first derivative of Eq. (7.1) over x is equivalent to multiplying the wave function $\Psi(x, t)$ by ik.
(2) Taking the first derivative of Eq. (7.1) over t is equivalent to multiplying the wave function $\Psi(x, t)$ by $-i\omega$.

Thus, from Eq. (7.3), we can conclude that the differential equation for the wave function should be the first order in t and the second order in x. The simplest equation of this form is

$$\frac{\partial \Psi(x, t)}{\partial t} = \Gamma^2 \frac{\partial^2 \Psi(x, t)}{\partial x^2} , \tag{7.4}$$

where Γ is a parameter, which has to be determined.

To determine Γ, we substitute Eq. (7.1) into Eq. (7.4) and find

$$-i\omega = -\Gamma^2 k^2 . \tag{7.5}$$

Then, using Eq. (7.3), we find that

$$\Gamma^2 = \frac{i\hbar}{2m} . \tag{7.6}$$

Hence, the wave function of a particle of the energy (7.2) satisfies the following differential equation:

$$\frac{\partial \Psi(x, t)}{\partial t} = \frac{i\hbar}{2m} \frac{\partial^2 \Psi(x, t)}{\partial x^2} , \tag{7.7}$$

or equivalently

$$i\hbar \frac{\partial \Psi(x, t)}{\partial t} + \frac{\hbar^2}{2m} \frac{\partial^2 \Psi(x, t)}{\partial x^2} = 0 . \tag{7.8}$$

Equation (7.8) is called the **one-dimensional Schrödinger equation** for a free particle. It is easy to extend the equation to three dimensions:

$$i\hbar \frac{\partial \Psi(\vec{r}, t)}{\partial t} + \frac{\hbar^2}{2m} \nabla^2 \Psi(\vec{r}, t) = 0 . \tag{7.9}$$

7.1.1 *Operators*

We can write the three-dimensional Schrödinger equation in the following form:

$$-\frac{\hbar}{i} \frac{\partial}{\partial t} \Psi(\vec{r}, t) = \frac{1}{2m} \left(\frac{\hbar}{i} \nabla \right) \left(\frac{\hbar}{i} \nabla \right) \Psi(\vec{r}, t) . \tag{7.10}$$

It shows that the Schrödinger equation can be obtained from the energy (Hamiltonian) of the free particle ($E = |\vec{p}|^2/2m$) by simply replacing E and $\vec{p}$, respectively, by

$$E \to -\frac{\hbar}{i} \frac{\partial}{\partial t} , \qquad \vec{p} \to \frac{\hbar}{i} \nabla . \tag{7.11}$$

These relations show that in quantum physics, the physical quantities are represented by mathematical operations. We call them **operators**.

The quantities ∇ and $\partial/\partial t$ define *operations* or *actions* to be carried out on the wave function Ψ. The particular operation stated in Eq. (7.10), $\partial \Psi/\partial t$ consists of taking a partial derivative of Ψ in terms of t, and $\nabla^2 \Psi$ consists of taking partial derivatives of Ψ in terms of Cartesian coordinates. The result is a new wave function, which may be different from the original one or may be equal to the original wave function multiplied by a scalar. We often say that operators associate a wave function with another wave function. To clarify further the action of the operators ∇ and $\partial/\partial t$ on the wave function, we will apply them to a specific example.

Worked Example

Calculate (a) $\frac{\partial \Psi}{\partial t}$ and (b) $\nabla^2 \Psi$, where $\Psi = \Psi_{max} e^{i(kx-\omega t)}$.

Solution

(a) The partial derivative of Ψ in terms of t is

$$\frac{\partial}{\partial t}\Psi = \Psi_{max}\frac{\partial}{\partial t}e^{i(kx-\omega t)} = -i\omega\Psi_{max}e^{i(kx-\omega t)} = -i\omega\Psi. \quad (7.12)$$

Thus, the action of the operator $\partial/\partial t$ on the wave function Ψ results in a constant $-i\omega$ times the original wave function. We shall see later that such a wave function is called in quantum physics an *eigenfunction* or *eigenstate* of the $\partial/\partial t$ operator, and $-i\omega$ is the corresponding *eigenvalue*.

The solution to part (b) is left to the readers.

Important property of operators

In classical physics, the multiplication of two quantities, say x and p_x, is immaterial. However, in quantum physics, where physical quantities are represented by operators, the order of multiplication is important and, for example, $xp_x \neq p_x x$, where $p_x = -i\hbar\partial/\partial x$. We say that the two quantities x and p_x do not commute.[a] The readers are familiar with such an ordering through the use of matrix algebra, where in general the order of two matrices is important, that is $M_1 M_2 \neq M_2 M_1$.

A measure of the extent to which $xp_x \neq p_x x$ is given by the *commutator* bracket

$$[\hat{x}, \hat{p}_x] = \hat{x}\hat{p}_x - \hat{p}_x\hat{x}, \quad (7.13)$$

where we have introduced the symbol "^" over the quantities x and p_x to indicate that these quantities are operators.

---

[a]Commutation consists in reversing the order of two quantities in an algebraic operation.

Note that the coordinates of $\vec{r}$ are the same in operator and classical forms. For example, the coordinate x is simply used in the operator form as x.

How to calculate the commutator $[\hat{x}, \hat{p}_x]$?

Since operators are "action" operations on functions, we consider the action of this commutator on a trial function $\Psi(x)$:

$$[\hat{x}, \hat{p}_x] \Psi(x) = x \left(-i\hbar \frac{\partial \Psi}{\partial x}\right) + i\hbar \frac{\partial}{\partial x} (x\Psi)$$

$$= -i\hbar x \frac{\partial \Psi}{\partial x} + i\hbar \Psi + i\hbar x \frac{\partial \Psi}{\partial x} = i\hbar \Psi . \quad (7.14)$$

Hence

$$[\hat{x}, \hat{p}_x] = i\hbar . \quad (7.15)$$

The result of the commutator is a number $i\hbar$. However, this is not the general rule that a commutator of two operators is always a number. We shall see later many examples where the commutator of two operators is an operator.

We can generalize the commutation relation between the position and momentum operators into three dimensions and can readily show that the components of the positions $\hat{r}$ and momentum $\hat{p}$ operators satisfy the commutation relations

$$[\hat{r}_m, \hat{p}_n] = i\hbar \delta_{mn} , \qquad m, n = 1, 2, 3 , \quad (7.16)$$

where

$$\hat{r}_1 = \hat{x} , \quad \hat{r}_2 = \hat{y} , \quad \hat{r}_3 = \hat{z} ,$$
$$\hat{p}_1 = \hat{p}_x , \quad \hat{p}_2 = \hat{p}_y , \quad \hat{p}_3 = \hat{p}_z . \quad (7.17)$$

The symbol δ_{mn} is called *Kronecker δ function* and is defined as

$$\delta_{mn} = \begin{cases} 1 & \text{if } m = n \\ 0 & \text{if } m \neq n . \end{cases} \quad (7.18)$$

The commutation relations (7.16) are called the *canonical commutation relations*.

Using the operator representation, the Schrödinger equation is often written as

$$i\hbar \frac{\partial \Psi(\vec{r}, t)}{\partial t} = \hat{H} \Psi(\vec{r}, t) , \quad (7.19)$$

where $\hat{H} = -\frac{\hbar^2}{2m} \nabla^2$ is the Hamiltonian (energy operator) of the free particle.

7.2 Schrödinger Equation of a Particle in an External Potential

In physics, we often deal with problems in which particles are free within some kind of boundary but have boundary conditions set by some external potentials. The particle-in-a-box problem, discussed before, is the simplest example.

In the presence of an external potential $V(\vec{r}, t)$, which may depend on time, the Hamiltonian of the particle takes the form

$$\hat{H}(\vec{r}, t) = -\frac{\hbar^2}{2m}\nabla^2 + \hat{V}(\vec{r}, t), \qquad (7.20)$$

which shows that the particle can gain an energy due to the potential $\hat{V}(\vec{r}, t)$.

The wave function of the particle moving in the external potential can be different from that of the free particle. It can be found solving the Schrödinger equation (7.19) with the Hamiltonian (7.20). The solution, however, must satisfy the following conditions:

(1) The wave function must be determined and continuous at any point of the space $(\vec{r}, t)$.
(2) The wave function must vanish at infinity, i.e., $\Psi(\vec{r}, t) \to 0$ when $r \to \pm\infty$.

We will try to solve the Schrödinger equation assuming that the Hamiltonian $\hat{H}(\vec{r}, t)$ does not explicitly depend on time, i.e., $\hat{V}(\vec{r}, t) = \hat{V}(\vec{r})$. In this case, the Schrödinger equation contains two terms: one dependent on time t and the other dependent on $\vec{r}$, i.e.,

$$\left(i\hbar\frac{\partial}{\partial t} - \hat{H}(\vec{r})\right)\Psi(\vec{r}, t) = 0, \qquad (7.21)$$

where $\hat{H}$ depends solely on $\vec{r}$.

Since the time- and $\vec{r}$-dependent parts are separated, the solution to the Schrödinger equation will be in the form of a product of two functions $\phi(\vec{r})$ and $f(t)$:

$$\Psi(\vec{r}, t) = \phi(\vec{r}) f(t). \qquad (7.22)$$

Substituting this equation into the Schrödinger equation, we get

$$i\hbar\phi(\vec{r})\frac{df(t)}{dt} = f(t)\hat{H}(\vec{r})\phi(\vec{r}), \qquad (7.23)$$

which can be written as

$$i\hbar\frac{1}{f}\frac{df}{dt} = \frac{1}{\phi}\hat{H}\phi \,, \tag{7.24}$$

where $\hat{H} \equiv \hat{H}(\vec{r})$, $f \equiv f(t)$, and $\phi \equiv \phi(\vec{r})$.

The left-hand side of Eq. (7.24) is a function of only one variable t, whereas the right-hand side is a function of only the position $\vec{r}$, i.e., each side is independent of any changes in the other. Thus, both sides must be equal to a constant, say E:

$$i\hbar\frac{1}{f}\frac{df}{dt} = E \,, \tag{7.25}$$

$$\frac{1}{\phi}\hat{H}\phi = E \,. \tag{7.26}$$

We can easily solve Eq. (7.25), and the solution can be written directly as

$$f(t) = Ce^{-\frac{i}{\hbar}Et} \,, \tag{7.27}$$

where C is a constant.

The other part of the Schrödinger equation, Eq. (7.26), can be written as

$$\hat{H}\phi = E\phi \,, \tag{7.28}$$

which is called the stationary (time-independent) Schrödinger equation, or the **eigenvalue equation** for the Hamiltonian $\hat{H}$.

Hence, the complete solution to the Schrödinger equation is of the form

$$\Psi(\vec{r}, t) = C\phi(\vec{r})e^{-\frac{i}{\hbar}Et} \,, \tag{7.29}$$

that is, $\Psi(\vec{r}, t)$ is the product of a time-dependent function and a position-dependent function $\phi(\vec{r})$, which satisfies the eigenvalue equation (7.28).

Note that the solution (7.29) leads to the probability density

$$|\Psi(\vec{r}, t)|^2 = |C\phi(\vec{r})|^2 \,, \tag{7.30}$$

which is independent of time.

Thus, when the Hamiltonian of a particle is independent of time, the probability of finding the particle in an arbitrary point $\vec{r}$ is independent of time. Such a state (wave function) is called a **stationary state** of the particle.

The existence of stationary states has two very useful practical consequences. Physically, such states have a permanence in time, which allows their long-time experimental investigations. Mathematically, they reduce the Schrödinger equation to the eigenvalue equation for the Hamiltonian. Thus, to obtain specific values of energy and corresponding wave functions, we operate on the wave function with the Hamiltonian and solve the resulting differential equation. However, not all mathematically possible solutions are accepted. Physics imposes some limits on the solutions of the Schrödinger equation.

More precisely, the solution to the Schrödinger equation must satisfy the following conditions:

(1) The wave function ϕ must be finite in all points of the space and vanish at infinity.
(2) The wave function ϕ must be continuous and should have continuous first derivatives.
(3) The wave function ϕ must be a single-value function at any point $\vec{r}$.
(4) The wave function must be normalized.

These conditions are often called the boundary conditions for the wave function.

When an operation on a wave function gives a constant times the original wave function, that constant is called an **eigenvalue** and the wave function is called an **eigenfunction** or **eigenstate**. Thus, the wave function that satisfies the stationary Schrödinger equation is the *eigenfunction* of the Hamiltonian $\hat{H}$, and E is the *eigenvalue* of the Hamiltonian in the state ϕ.

The complete set of eigenvalues of the Hamiltonian $\hat{H}$ is termed **energy spectrum**. The energy spectrum can be non-degenerated (different eigenfunctions have different eigenvalues), or degenerated (all or few eigenfunctions have the same eigenvalues), but it is not allowed that one eigenfunction could have a few different eigenvalues.

To solve the Schrödinger equation, even in its simpler stationary (time-independent) form, usually requires sophisticated mathematical techniques. In fact, for any system, the stationary Schrödinger equation

$$\hat{H}\phi = E\phi \qquad (7.31)$$

is in the form of a second-order differential equation

$$-\frac{\hbar^2}{2m}\nabla^2\phi + \left(\hat{V}(\vec{r}) - E\right)\phi = 0, \qquad (7.32)$$

whose solution depends on the explicit form of the potential $\hat{V}(\vec{r})$. Thus, we see that the particle is represented by its mass and only the potential $\hat{V}(\vec{r})$ alters the form of the differential equation.

Hence, the problem of finding energies and the wave function of a particle moving in a potential $\hat{V}(\vec{r})$ is equivalent to solving the second-order differential equation for the wave function with a known specific form of $\hat{V}(\vec{r})$. In other words, once $\hat{V}(\vec{r})$ is known, the Schrödinger equation may be solved and the wave function $\phi(\vec{r})$ may be found. Then the probability density $|\phi(\vec{r})|^2$ may be determined for a specific point $\vec{r}$.

In the next few chapters, we will investigate solutions to the stationary Schrödinger equation for different forms of the potential $\hat{V}(\vec{r})$. More precisely, we will seek energies and wave functions of a particle subject to different constrains on its free motion.

In summary of the chapter on the Schrödinger equation, we have learned that

(1) In quantum physics, physical quantities are represented by operators.
(2) The operator representing the energy of a system is the Hamiltonian $\hat{H}$.
(3) The eigenvalues of $\hat{H}$ are energies E.
(4) If the potential $\hat{V}$ is independent of time, then the separation of variables is possible and we can write the wave function as $\Psi(\vec{r}, t) = \phi(\vec{r})f(t)$.
(5) The wave function $\phi(\vec{r})$ is the eigenfunction of the time-independent Hamiltonian $\hat{H}$ and can be found by solving the stationary Schrödinger equation $\hat{H}\phi(\vec{r}) = E\phi(\vec{r})$.

7.3 Equation of Continuity

We know that the probability of finding a particle in a volume V is normalized to 1, i.e.,

$$\int_V |\Psi(\vec{r}, t)|^2 dV = 1. \tag{7.33}$$

This normalization condition must be valid for any wave function evaluated at any point $\vec{r}$ and at any time t. We will show that the Schrödinger equation guarantees the conservation of normalization of the wave function. In other words, if Ψ was normalized at $t = 0$, it will remain normalized at all times.

In addition, there is a flow of the probability density or particle current density associated with a moving particle. Therefore, we will also define what the particle probability current density is in terms of the particle wave function.

Suppose we have a particle described by a wave function Ψ in a volume V enclosed by a surface S. We will consider the time evolution of the particle wave function, which is given by the time-dependent Schrödinger equation

$$i\hbar \frac{\partial \Psi}{\partial t} = -\frac{\hbar^2}{2m} \nabla^2 \Psi + \hat{V} \Psi . \tag{7.34}$$

First, we take complex conjugate of the aforementioned equation:

$$-i\hbar \frac{\partial \Psi^*}{\partial t} = -\frac{\hbar^2}{2m} \nabla^2 \Psi^* + \hat{V} \Psi^* . \tag{7.35}$$

Next, multiplying Eq. (7.34) by Ψ^* and Eq. (7.35) by Ψ, and subtracting the resulting equations, we get

$$i\hbar \left(\Psi^* \frac{\partial \Psi}{\partial t} + \Psi \frac{\partial \Psi^*}{\partial t} \right) = -\frac{\hbar^2}{2m} \left(\Psi^* \nabla^2 \Psi - \Psi \nabla^2 \Psi^* \right) . \tag{7.36}$$

Note that

$$\Psi^* \frac{\partial \Psi}{\partial t} + \Psi \frac{\partial \Psi^*}{\partial t} = \frac{\partial}{\partial t} |\Psi|^2 . \tag{7.37}$$

Moreover, using a vector identity

$$\nabla \cdot \left(u\vec{A} \right) = \nabla u \cdot \vec{A} + u\nabla \cdot \vec{A} , \tag{7.38}$$

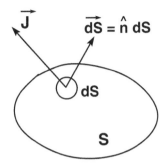

Figure 7.1 Probability current density $\vec{J}$ crossing a surface S, with $\hat{n}$, the unit vector normal to the surface.

we find that

$$\nabla \cdot (\Psi^* \nabla \Psi - \Psi \nabla \Psi^*)$$
$$= \nabla \Psi^* \cdot \nabla \Psi + \Psi^* \nabla \cdot (\nabla \Psi) - \nabla \Psi \cdot \nabla \Psi^* - \Psi \nabla \cdot (\nabla \Psi^*)$$
$$= \Psi^* \nabla^2 \Psi - \Psi \nabla^2 \Psi^* . \tag{7.39}$$

Thus, Eq. (7.36) can be written as

$$\frac{\partial}{\partial t} |\Psi|^2 + \nabla \cdot \left(\frac{\hbar}{2im} (\Psi^* \nabla \Psi - \Psi \nabla \Psi^*) \right) = 0 . \tag{7.40}$$

Introducing a notation

$$|\Psi|^2 = \rho , \qquad \frac{\hbar}{2im} (\Psi^* \nabla \Psi - \Psi \nabla \Psi^*) = \vec{J} , \tag{7.41}$$

we obtain

$$\frac{\partial \rho}{\partial t} + \nabla \cdot \vec{J} = 0 . \tag{7.42}$$

The reader familiar with the theory of fluids and with electricity and magnetism will immediately recognize that the above equation is identical in form to the well-known **continuity equation**, which shows the conservation of matter or the conservation of charge. In our case, the continuity equation shows the conservation of the probability density ρ, and $\vec{J}$ is then the probability current density. To interpret the continuity equation in terms of a flow of the probability, it is convenient to integrate Eq. (7.42) over the volume V closed by a surface S, as shown in Fig. 7.1:

$$\frac{\partial}{\partial t} \int_V |\Psi|^2 dV = - \int_V \nabla \cdot \vec{J} dV . \tag{7.43}$$

From Gauss's divergence theorem

$$\int_V \nabla \cdot \vec{J} \, dV = \oint_S \vec{J} \cdot d\vec{S}, \qquad (7.44)$$

we get

$$\frac{\partial}{\partial t} \int_V |\Psi|^2 dV = -\oint_S \vec{J} \cdot d\vec{S}. \qquad (7.45)$$

The left-hand side of this equation is the time rate of increase of probability of finding the particle inside the volume V. The integral on the right-hand side is the probability per unit time of the particle leaving the volume V through the surface S. This justifies why $\vec{J}$ is called the probability current density. It tells us the rate at which probability is "flowing" through the surface S.

The scalar product $\vec{J} \cdot d\vec{S}$ is the probability that the particle will cross an area $d\vec{S}$ on the surface. When the particle remains inside the volume for all times, i.e., does not cross the surface $\vec{S}$, then $\vec{J} \cdot d\vec{S} = 0$, and we get

$$\frac{\partial}{\partial t} \int_V |\Psi|^2 dV = 0, \qquad (7.46)$$

which shows that the Schrödinger equation guarantees the conservation of normalization of the wave function. In other words, if Ψ was normalized at $t = 0$, it will remain normalized at all times.

7.4 Transmission and Reflection Coefficients

Suppose that particles inside the surface are represented by plane waves

$$\Psi_{\text{in}}(\vec{r}) = A e^{i\vec{k}_1 \cdot \vec{r}} + B e^{-i\vec{k}_1 \cdot \vec{r}}, \qquad (7.47)$$

and outside the surface

$$\Psi_{\text{out}}(\vec{r}) = C e^{i\vec{k}_2 \cdot \vec{r}}, \qquad (7.48)$$

where $\vec{k}_1$ and $\vec{k}_2$ are the wave vectors of the particle inside and outside the surface, respectively. The coefficients A, B, and C are interpreted as the amplitudes of the incident, reflected, and transmitted particles, respectively.

To interpret the wave functions, we calculate the probability current densities inside and outside the surface and find

$$\vec{J}_{\text{in}} = \frac{\hbar}{2im} \left(\Psi_{\text{in}}^* \nabla \Psi_{\text{in}} - \Psi_{\text{in}} \nabla \Psi_{\text{in}}^* \right) = \frac{\hbar \vec{k}_1}{m} \left(|A|^2 - |B|^2 \right), \quad (7.49)$$

and

$$\vec{J}_{\text{out}} = \frac{\hbar}{2im} \left(\Psi_{\text{out}}^* \nabla \Psi_{\text{out}} - \Psi_{\text{out}} \nabla \Psi_{\text{out}}^* \right) = \frac{\hbar \vec{k}_2}{m} |C|^2. \quad (7.50)$$

Inside the surface, the current density is composed of two terms and can be written as

$$\vec{J}_{\text{in}} = \vec{J}_i - \vec{J}_r, \quad (7.51)$$

where

$$\vec{J}_i = \frac{\hbar \vec{k}_1}{m} |A|^2 \quad (7.52)$$

is interpreted as the **incident particle current**, and

$$\vec{J}_r = \frac{\hbar \vec{k}_1}{m} |B|^2 \quad (7.53)$$

is interpreted as the **reflected particle current**. The current density outside the surface

$$\vec{J}_t = \frac{\hbar \vec{k}_2}{m} |C|^2 \quad (7.54)$$

is interpreted as the **transmitted particle current**.

If the number of particles is conserved, i.e., the particles are not created or destroyed, the current densities inside and outside the surface should be equal:

$$\vec{J}_{\text{in}} = \vec{J}_{\text{out}}. \quad (7.55)$$

This equation can be written in terms of the amplitudes as

$$|A|^2 - |B|^2 = \frac{k_2}{k_1} |C|^2. \quad (7.56)$$

We can define the **reflection coefficient** of the surface

$$R = \frac{|\vec{J}_r|}{|\vec{J}_i|}, \quad (7.57)$$

which is given by the probability current density reflected from the surface divided by the probability current density incident on the surface.

We can also define the **transmission coefficient** of the surface

$$T = \frac{|\vec{J}_t|}{|\vec{J}_i|}, \tag{7.58}$$

which is given by the probability current density transmitted through the surface divided by the probability current density incident on the surface.

Using Eqs. (7.52)–(7.54), we can write the reflection and transmission coefficients in terms of the incident, reflected, and transmitted coefficients as

$$R = \frac{|B|^2}{|A|^2}, \qquad T = \frac{k_2}{k_1} \frac{|C|^2}{|A|^2}, \tag{7.59}$$

where $k_1 = |\vec{k}_1|$ and $k_2 = |\vec{k}_2|$.

Thus, if $|B|^2 = |A|^2$, then $R = 1$, i.e., all the particles that are incident on the surface are reflected.

It is easy to show from Eqs. (7.56) and (7.59) that

$$R + T = 1, \tag{7.60}$$

as it should be, otherwise the particles could be created or destroyed.

Revision Questions

Question 1 What are the conditions imposed on the wave function that is a solution to the time-independent Schrödinger equation?

Question 2 Define the probability current density and what does it describe?

Question 3 Define the transmission coefficient from the region where a particle has a momentum k_1 to a region where the particle has a momentum k_2.

Discussion Problem

Problem D2 One can notice from the definition of the probability current density, Eq. (7.41), that in general when the wave function

Ψ of a particle in a given region is real, the current density $\vec{J} = 0$ in this region.

How would you interpret this result?

Tutorial Problems

Problem 7.1 Usually we find the wave function from knowing the potential $V(x)$. Consider, however, an inverse problem where we know the wave function and would like to determine the potential that leads to the behavior described by the wave function.

Assume that a particle is confined within the region $0 \leq x \leq a$, and its the wave function is

$$\phi(x) = \sin\left(\frac{\pi x}{a}\right).$$

Using the stationary Schrödinger equation, find the potential $V(x)$ confining the particle.

Problem 7.2 Another example of the inverse problem where we know the wave function and would like to determine the potential that leads to the behavior described by the wave function.

Let $\phi(x)$ be the one-dimensional stationary wave function

$$\phi(x) = A\left(\frac{x}{x_0}\right)^n e^{-x/x_0},$$

where A, x_0, and n are constants.

Using the stationary Schrödinger equation, find the potential $V(x)$ and the energy E for which this wave function is an eigenfunction.

Assume that $V(x) \to 0$ as $x \to \infty$.

Problem 7.3 Consider the three-dimensional time-dependent Schrödinger equation of a particle of mass m moving in a potential $\hat{V}(\vec{r}, t)$:

$$i\hbar \frac{\partial \Psi(\vec{r}, t)}{\partial t} = \left(-\frac{\hbar^2}{2m}\nabla^2 + \hat{V}(\vec{r}, t)\right)\Psi(\vec{r}, t).$$

(a) Explain, what must be assumed about the form of the potential energy to make the equation separable into a time-independent

Schrödinger equation and an equation for the time dependence of the wave function.

(b) Using the condition stated in **(a)**, separate the time-dependent Schrödinger equation into a time-independent Schrödinger equation and an equation for the time-dependent part of the wave function.

(c) Solve the equation for the time-dependent part of the wave function and explain why the wave function of the separable Schrödinger equation is a stationary state of the particle.

Problem 7.4 Consider the wave function

$$\Psi(x, t) = \left(Ae^{ikx} + Be^{-ikx}\right) e^{i\omega t} .$$

(a) Find the probability current corresponding to this wave function.

(b) How would you interpret the physical meaning of the parameters A and B?

Chapter 8

Applications of Schrödinger Equation: Potential (Quantum) Wells

We have seen that the wave nature of particles plays an important role in their physical properties that, for example, particles confined into a small bounded area can have only particular discrete energies. However, the question we are most interested in is: Can we create an artificial structure that exploits discrete energy levels? The answer is "yes"; we can produce such structures, and they involve potential barriers. Such one-dimensional structures constructed are called **quantum wells**, two-dimensional structures are called **quantum wires**, and three-dimensional structures are called **quantum dots**.

With the current knowledge of quantum physics, this answer may probably sound rather abstract to the readers, so let us try to make it more concrete.

To illustrate that in practice, particles may really exhibit unusual quantum effects when they are located in such structures; we will solve the time-independent Schrödinger equation

$$\hat{H}\phi(\vec{r}) = E\phi(\vec{r}), \tag{8.1}$$

to find the energies (eigenvalues) E and the corresponding eigenfunctions $\phi(\vec{r})$ of a particle of mass m moving in a potential $\hat{V}(\vec{r})$ that varies with the position $\vec{r}$.

Quantum Physics for Beginners
Zbigniew Ficek
Copyright © 2016 Pan Stanford Publishing Pte. Ltd.
ISBN 978-981-4669-38-2 (Hardcover), 978-981-4669-39-9 (eBook)
www.panstanford.com

It is simplest to first consider the motion of particles in one dimension. Therefore, in this and the following chapter, we will limit our calculations to the *one-dimensional case*, in which the Hamiltonian of the particle moving in one dimension, say x-direction, is given by

$$\hat{H} = -\frac{\hbar^2}{2m}\frac{d^2}{dx^2} + \hat{V}(x). \tag{8.2}$$

With this Hamiltonian, we get from the Schrödinger equation a second-order differential equation for the wave function of the particle

$$\frac{d^2\phi(x)}{dx^2} = -\frac{2m}{\hbar^2}(E - V(x))\phi(x), \tag{8.3}$$

which can be written as

$$\frac{d^2\phi(x)}{dx^2} = -k^2(x)\phi(x), \tag{8.4}$$

where

$$k^2(x) = \frac{2m\left(E - \hat{V}(x)\right)}{\hbar^2}. \tag{8.5}$$

Thus, we see that the behavior of the particle, which is determined by the parameter $k^2(x)$, will depend *only* on three factors: total energy of the particle, potential barriers, and the mass of the particle.

When $\hat{V}(x)$ is independent of x, i.e., the particle is moving along the x-axis under the influence of no force because the potential is constant, the parameter $k^2(x) = k^2$, and then Eq. (8.4) reduces to a simple harmonic oscillator equation

$$\frac{d^2\phi(x)}{dx^2} = -k^2\phi(x). \tag{8.6}$$

This is a linear differential equation with a constant coefficient. The solution to Eq. (8.6) depends on whether $k^2 > 0$ or $k^2 < 0$.

For $k^2 > 0$, the general solution to Eq. (8.6) is in the form of an oscillating wave

$$\phi(x) = Ae^{ikx} + Be^{-ikx}, \qquad (E > V), \tag{8.7}$$

where A and B are amplitudes of the particle wave moving to the right and to the left, respectively.

For $k^2 < 0$, the general solution to Eq. (8.6) is in the form

$$\phi(x) = Ce^{-kx} + De^{kx}, \qquad (E < V), \qquad (8.8)$$

that the exponents are real and no longer represent an oscillating wave function. They represent a wave function with damped amplitudes.

Important note: The general solution (8.7) with both constants A and B different from zero is physically acceptable. However, the general solution (8.8) with both constants C and D different from zero cannot be accepted. We have learned that the wave function must vanish for $x \to \pm\infty$. Thus, if the particle moves in the direction of positive x, in an unbounded or semibounded space, then only the wave function with $C \neq 0$ and $D = 0$ will satisfy this condition, whereas if the particle moves in the direction of negative x, only the wave function with $C = 0$ and $D \neq 0$ will satisfy the condition of $\phi(x) \to 0$ as $x \to -\infty$.

Another important observation: The general solutions (8.7) and (8.8) are single-value solutions for the wave function $\phi(x)$. Thus, for the particle moving in an unbounded area where the potential $\hat{V}$ is constant, there are no restrictions on k, which, according to Eq. (8.5), means that there are no restrictions on the energy E of the particle. Hence, the energy E of the particle can have any value ranging from zero to $+\infty$ (continuous spectrum). It is also valid for x-dependent potentials, where $V(x)$ slowly changes with x.[a]

Having now obtained a general solution to the Schrödinger equation, let us examine a few special cases more closely. We will consider four cases of a one-dimensional motion of particles confined in potentials rapidly changing with x:

- Infinite potential quantum well.
- A potential step.
- Square-well potential.
- Tunneling through a potential barrier.

---

[a] For potentials rapidly changing with x, the particle can be trapped in potential holes, and then E can be different.

Our interest in these four problems is

(1) To understand how the wave function of a particle confined by a potential V is calculated.
(2) To see how the Schrödinger equation is solved when the motion of a particle is subject to restrictions.
(3) To learn the characteristic properties of solutions to this equation.
(4) To see differences between the predictions of quantum mechanics and classical physics.

8.1 Infinite Potential Quantum Well

As the first example of the application of Schrödinger equation, consider a particle confined in a one-dimensional structure, an infinite potential well, as illustrated in Fig. 8.1. The term "well" is a bit misleading since the particle is actually only trapped in one direction. It is still free to move in other two directions. However, the term "well" is commonly used in the literature and we will follow this terminology.

For the infinite potential well centered at $x = 0$:

$$V(x) = 0 \qquad \text{for} \qquad -\frac{a}{2} \leq x \leq \frac{a}{2},$$

$$V(x) = \infty \qquad \text{for} \qquad x < -\frac{a}{2} \quad \text{and} \quad x > \frac{a}{2}. \qquad (8.9)$$

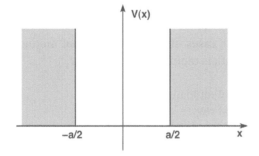

Figure 8.1 An infinite potential well. Outside the region $-a/2 \leq x \leq a/2$, the potential $V(x) \to \infty$.

Within the well, there is no potential energy, while outside the well the potential is infinite, so that the particle cannot exist there since it would have to have infinite energy. What classical physics and quantum physics tell us about the behavior of the particle inside the well?

According to classical physics, the particle trapped between the potential walls will bounce back and forth indefinitely; its kinetic energy will be constant $E = mv^2/2$. Moreover, the probability of finding the particle at any point between the walls is constant and anywhere outside the walls is zero. In fact, if we know the initial momentum and position of the particle, we can specify the location of the particle at any time in the future. The classical case seems trivial.

According to quantum physics, the particle is described by a wave function $\phi(x)$, which satisfies the Schrödinger equation and some boundary conditions. One of the boundary conditions says that the wave function $\phi(x)$ must be finite everywhere. Thus, in the regions $x < -a/2$ and $x > a/2$, the wave function $\phi(x)$ must be zero to satisfy this condition that $V(x)\phi(x)$ must be finite everywhere.

In the region $-a/2 \leq x \leq a/2$, the potential $V(x) = 0$, and then the Schrödinger equation for the wave function takes the form

$$\frac{d^2\phi(x)}{dx^2} = -k^2\phi(x),\qquad(8.10)$$

where $k^2 = 2mE/\hbar^2$.

Since k^2 is positive, the Schrödinger equation (8.10) has a simple solution

$$\phi(x) = Ae^{ikx} + Be^{-ikx},\qquad -\frac{a}{2} \leq x \leq \frac{a}{2},\qquad(8.11)$$

where A and B are constants, which in general are complex numbers.

To determine the unknown constants,[a] we will use the boundary condition that the wave function must be continuous at $x = -a/2$ and $x = a/2$.

---

[a]Usually, we find only one of the two constants in terms of the other, say B in terms of A. The remaining constant A is readily found from the normalization condition that the wave function is normalized to 1, i.e.,

$$\int_{-\infty}^{\infty} dx\, |\phi(x)|^2 = 1.\qquad(8.12)$$

Since in the regions $x < -a/2$ and $x > a/2$, the wave function is equal to zero and the wave function must be continuous at $x = -a/2$ and $x = a/2$, we have $\phi(x) = 0$ at these points. In other words, the wave functions must join smoothly at these points.

Thus, at $x = -a/2$, the wave function $\phi(x) = 0$ when

$$Ae^{-\frac{ika}{2}} + Be^{\frac{ika}{2}} = 0 . \tag{8.13}$$

At $x = a/2$, the wave function $\phi(x) = 0$ when

$$Ae^{\frac{ika}{2}} + Be^{-\frac{ika}{2}} = 0 . \tag{8.14}$$

From Eq. (8.13), we find that

$$B = -Ae^{-ika} , \tag{8.15}$$

whereas from Eq. (8.14), we find that

$$B = -Ae^{ika} . \tag{8.16}$$

We have obtained two different solutions for the coefficient B. Accepting these two different solutions, we would accept two different solutions to the wave function. However, we cannot accept it, as one of the conditions imposed on the wave function says that the wave function must be a single-value function. Therefore, we have to find a condition under which the two solutions (8.15) and (8.16) are equal. It is easy to see from Eqs. (8.15) and (8.16) that the two solutions for B will be equal if

$$e^{-ika} = e^{ika} , \tag{8.17}$$

which will be satisfied when

$$e^{2ika} = \cos(2ka) + i\sin(2ka) = 1 , \tag{8.18}$$

or when

$$\sin(2ka) = 0 \quad \text{and} \quad \cos(2ka) = 1 , \tag{8.19}$$

i.e., when

$$k = n\frac{\pi}{a} , \quad \text{with} \quad n = 0, 1, 2, \dots . \tag{8.20}$$

Thus, for a particle confined in the infinite well, a restriction is imposed on k that k can take only discrete values.

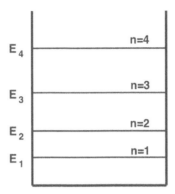

Figure 8.2 Four lowest energy levels of a particle trapped inside the infinite potential well. Note that the separation between the energy levels increases with an increasing n.

8.1.1 *Energy Quantization*

Since $k^2 = 2mE/\hbar^2$, and k is restricted to discrete values, we see that the energy of the particle cannot be arbitrary; it can take only certain discrete values!

$$E_n = \frac{\hbar^2}{2m}k^2 = n^2 \frac{\pi^2 \hbar^2}{2ma^2}. \tag{8.21}$$

Thus, the energy of the particle inside the well is quantized and can have only discrete values (discrete spectrum), which depend on the integer variable n.

We indicate this by writing a subscript n on E. The integer number n is called the **quantum number**. A few of the lowest energy levels are shown in Fig. 8.2. Note from Eq. (8.21) that the energy levels in a quantum well depend on the dimension of the well and the mass of the particle. This means that we can build artificial structures of desired quantum properties, which could be observed if the dimensions of the structures are very small.

8.1.2 *Wave Functions*

Substituting one of the solutions for B, Eq. (8.15) or (8.16) into the general solution (8.11), we find the wave function of the particle inside the well

$$\phi_n(x) = A' \sin\left[\frac{n\pi}{a}\left(x - \frac{a}{2}\right)\right], \quad \text{with} \quad n = 1, 2, 3, \ldots, \quad (8.22)$$

and $A' = A \exp(ika/2)$.

One may notice that the solution for $n = 0$ is not included. There is a simple explanation. For $n = 0$, the wave function $\phi(x) = 0$ for all x inside the well. Accepting this solution would mean we accept that the particle is not in the well. Thus, the minimum energy state in which the particle can be inside the well is that with the energy $E_1 = \pi^2\hbar^2/2ma^2$. Since $E_1 > 0$, the particle can never have zero energy. In other words, the particle can never truly be at rest.

This essentially solves the problem. The remaining coefficient A' that appears in Eq. (8.22) is found from the normalization condition

$$\int_{-\infty}^{+\infty} |\phi_n(x)|^2 dx = 1. \quad (8.23)$$

Performing integration with the wave function $\phi_n(x)$ given by Eq. (8.22), we find $|A'| = \sqrt{2/a}$. The details of the integration are left as an exercise for the readers.

While $\phi_n(x)$ may be negative as well as positive, $|\phi_n(x)|^2$ is always positive and, since $\phi_n(x)$ is normalized, its value at a given x is equal to the probability density of finding the particle at this point. Figure 8.3 shows the wave function $\phi_n(x)$ of the particle for the first three values of n. At a given x, the wave function is different for different n. For example, $\phi_1(x)$ has its maximum at $x = 0$, while $\phi_2(x) = 0$ at this particular point. For all n's, the probability is not constant

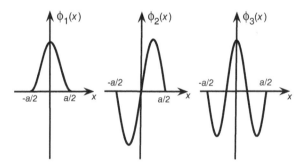

Figure 8.3 Plot of the wave function $\phi_n(x)$ for the first three energy levels $n = 1$, $n = 2$, and $n = 3$.

and, for $n > 1$, has zeros for some values of x. For example, in the lowest energy level of $n = 1$, the particle is most likely to be found in the center of the well, while being in the first excited state of $n = 2$, it will never be found there! This is in contrast to the predictions of classical physics, where the particle has the same probability of being located anywhere between the walls independent of its energy. Moreover, the lowest energy ($n = 1$) is nonzero, which indicates that the particle can have nonzero energy even if the potential energy is zero.

The predicted unusual effects: the exclusion of $E = 0$ as a possible value for the energy of the particle, the limitation of E to a discrete set of definite values, and the dependence of the probability on the position of the particle are other examples of quantum effects without classical analog.

In the next few chapters, we will learn that these unusual phenomena are not the only quantum phenomena. There is a more bizarre phenomenon: penetration of the barrier that particles can be found in the classically forbidden region.

What do we mean by the classically forbidden region? It is the energy region in which the particles, if found, would have to have *negative* kinetic energy.

In summary, we have learned that

(1) The energy of a particle in a quantum well can take only certain discrete values. All other values of the energy are forbidden. We say that the energy of the particle is quantized.
(2) The quantization of the energy arises from the condition of the continuity of the particle wave function at the boundaries between two regions of different potential.
(3) The lowest energy the particle can have inside the well is not zero.
(4) The probability of finding the particle at an arbitrary position x is not constant; it even has zeros.

Worked Example

An electron is confined in an infinite potential well of width $a = 0.1$ nm (approximate size of an atom).

(a) Calculate the minimum energy of the electron.
(b) What is the equivalent temperature?

Solution

(a) The minimum energy of the electron corresponds to $n = 1$. Using Eq. (8.21), we find

$$E_1 = (1)^2 \frac{\pi^2 \hbar^2}{2ma^2} = \frac{h^2}{8ma^2} = \frac{(6.626 \times 10^{-34})^2}{8 \times 9.109 \times 10^{-31} \times (10^{-10})^2}$$

$$= 6.025 \times 10^{-18} \text{ [J]} = 37.6 \text{ [eV]} .$$

(b) We find the temperature from the formula for the number of photons in the mode of frequency ω:

$$\langle n \rangle = \frac{1}{e^{\frac{E_n}{k_B T}} - 1} .$$

Since $n = 1$, we find that

$$e^{\frac{E_1}{k_B T}} = 2 ,$$

from which, we get

$$\frac{E_1}{k_B T} \approx 1 ,$$

i.e.,

$$T \approx E_1 / k_B = 6.025 \times 10^{-18} / \left(1.381 \times 10^{-13} \right) = 43.6 \text{ [μK]} .$$

This is a very low temperature, which is routinely achieved in laboratory with current trapping and cooling techniques of atomic gases.

Revision Questions

Question 1 Explain the reason for quantization of the energy of a particle confined in an infinite potential well.

Question 2 Explain why zero energy of the particle is excluded from the solution to the wave function of a particle confined inside an infinite potential well.

Question 3 Probability of finding the particle at any point inside the well is independent of the position of the particle. True or false?

Tutorial Problems

Problem 8.1 One may notice from Fig. 8.3 that the wave function for $n = 2$ is zero at $x = 0$, i.e., at the center of the well. This means that the probability of finding the particle at the center of the well is also zero. Then, a question arises: How does the particle move from one side of the well to the other if the probability of being at the center is zero?

Problem 8.2 Solve the stationary Schrödinger equation for a particle not bounded by any potential and show that its total energy E is not quantized.

Problem 8.3 Solve the Schrödinger equation with appropriate boundary conditions for an infinite square-well with the width of the well a centered at $a/2$, i.e.,

$$V(x) = 0 \qquad \text{for} \qquad 0 \le x \le a,$$
$$V(x) = \infty \qquad \text{for} \qquad x < 0 \quad \text{and} \quad x > a.$$

Check that the allowed energies are consistent with those derived in the chapter for an infinite well of width a centered at the origin. Confirm that the wave function $\phi_n(x)$ can be obtained from those found in chapter if one uses the substitution $x \to x + a/2$.

Problem 8.4 Show that, as $n \to \infty$, the probability of finding a particle between x and $x + \Delta x$ inside an infinite potential well is independent of x, which is the classical expectation. This result is an example of the correspondence principle that quantum theory should give the same results as classical physics in the limit of large quantum numbers.

Problem 8.5 As we have already learned, the exclusion of $E = 0$ as a possible value for the energy of the particle and the limitation of E to a discrete set of definite values are examples of quantum effects that have no counterpart in classical physics, where all energies, including zero, are presumed possible.

Why we do not observe these quantum effects in everyday life?

Problem 8.6 What length scale is required to observe discrete (quantized) energies of an electron confined in an infinite potential well?

Calculate the width of the potential well in which a low-energy electron, being in the energy state $n = 2$, emits a visible light of wavelength $\lambda = 700$ nm (red) when making a transition to its ground state $n = 1$. Compare the length scale (width) to the size of an atom ~ 0.1 nm.

8.2 Potential Step

In our studies of unusual properties of particles confined to a small region, we now remove one of the two barriers and make the other of finite potential V_0. This is called a potential step. Such a potential energy step very often appears in practice, e.g., at the connection of two different wires or semiconductors. Electrons moving inside either wire have a constant energy, but it changes very rapidly when passing from one to the other due to different conductivities of the wires.

Let us define the potential as

$$V(x) = V_0\Theta(x) \qquad \text{with} \qquad V_0 > 0, \qquad (8.24)$$

where $\Theta(x)$ is the Heaviside step function in which $\Theta(x) = 1$ for $x \geq 0$, and $\Theta(x) = 0$ for $x < 0$. This is shown in Fig. 8.4.

Thus, the potential has a height of V_0 for positive x and is zero for the negative x. This potential creates a barrier for particles moving along the x-axis.

An example of such a situation in practice is a junction between two conductors of different conductivities or between two different semiconductors. The junction works as a potential barrier for electrons on each side. Another practical example of a step potential is the surface of a conductor as the potential energy of an electron rapidly increases at the surface. Of course a step potential is an idealization. In real situations, potential does not change abruptly. Nevertheless, these idealized potentials are used frequently in quantum mechanics to approximate real situations. Because of the mathematical simplicity, one obtains the exact solution to the Schrödinger equation with a variety of initial conditions.

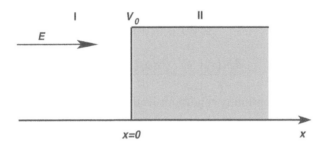

Figure 8.4 Potential step at $x = 0$. The potential is zero for $x < 0$ (region I) and $V = V_0$ for $x \geq 0$ (region II). Particles of total energy E travel from $-x$ toward the barrier.

Suppose that particles of mass m and total energy E travel from $-x$ toward the potential step (barrier), as shown in Fig. 8.4.

We will calculate the wave function of the particles in both regions $x < 0$ and $x \geq 0$ for two cases: $E < V_0$ and $E > V_0$.

The case $E < V_0$.

Since in region $x < 0$ (we will call it region I), the potential $V = 0$, the parameter k^2 appearing in the stationary Schrödinger equation is a positive number and, therefore, the solutions to the Schrödinger equation in this region are of the form

$$\text{I.} \quad \phi_1(x) = Ae^{ik_1x} + Be^{-ik_1x}, \qquad x < 0, \qquad (8.25)$$

where $k_1 = \sqrt{2mE}/\hbar$.

In region $x \geq 0$ (we will call it region II), the potential $V = V_0$. Since $E < V_0$, the parameter k^2 is a negative number and, therefore, the solutions to the Schrödinger equation in this region are of the form

$$\text{II.} \quad \phi_2(x) = C e^{k_2 x} + D e^{-k_2 x}, \qquad x \geq 0, \tag{8.26}$$

where $k_2 = \sqrt{2m(V_0 - E)}/\hbar$.

Since the wave function must be finite everywhere and must vanish as x becomes infinite, the coefficient C must be zero. Otherwise, $\phi_2(x)$ would go to infinity as $x \to \infty$.

We will find the coefficients B and D in terms of A, using the continuity conditions for the wave function, that at $x = 0$

$$\phi_1(0) = \phi_2(0),$$

$$\left. \frac{d\phi_1(x)}{dx} \right|_{x=0} = \left. \frac{d\phi_2(x)}{dx} \right|_{x=0}. \tag{8.27}$$

The above continuity conditions lead to two coupled equations:

$$A + B = D,$$
$$i k_1 A - i k_1 B = -k_2 D, \tag{8.28}$$

which can be written as

$$A + B = D,$$
$$A - B = i\beta D, \tag{8.29}$$

where $\beta = k_2/k_1$. By adding these two equations, we obtain

$$2A = (1 + i\beta)D, \tag{8.30}$$

and by subtracting the equations, we find

$$2B = (1 - i\beta)D. \tag{8.31}$$

Thus, we find that the coefficients B and D are given in terms of A as

$$D = \frac{2A}{(1 + i\beta)},$$

$$B = \frac{(1 - i\beta)}{(1 + i\beta)} A = \frac{(1 - i\beta)^2}{1 + \beta^2} A. \tag{8.32}$$

Hence, the wave function of the particle with $E < V_0$ is given by

$$\phi_1(x) = A e^{i k_1 x} + \frac{(1 - i\beta)^2}{1 + \beta^2} A e^{-i k_1 x}, \tag{8.33}$$

$$\phi_2(x) = \frac{2A}{(1 + i\beta)} e^{-k_2 x}. \tag{8.34}$$

The coefficient A can be found from the normalization condition.

Having the wave functions available, we can show that in both regions, $x < 0$ and $x > 0$, the probability current $\vec{J} = 0$.

The probability current is defined as

$$\vec{J} = \frac{\hbar}{2im}\left(\phi^* \frac{d\phi}{dx} - \phi \frac{d\phi^*}{dx}\right). \tag{8.35}$$

In region I, the wave function of the particles is

$$\phi_1(x) = A e^{ik_1 x} + \gamma A e^{-ik_1 x}, \tag{8.36}$$

where

$$\gamma = \frac{(1 - i\beta)^2}{1 + \beta^2}. \tag{8.37}$$

Hence

$$\frac{d\phi_1}{dx} = ik_1 A \left(e^{ik_1 x} - \gamma e^{-ik_1 x}\right), \tag{8.38}$$

and then

$$\phi_1^* \frac{d\phi_1}{dx} = ik_1|A|^2 \left(e^{-ik_1 x} + \gamma^* e^{ik_1 x}\right)\left(e^{ik_1 x} - \gamma e^{-ik_1 x}\right)$$

$$= ik_1|A|^2 \left[1 - |\gamma|^2 + \left(\gamma^* e^{2ik_1 x} - \gamma e^{-2ik_1 x}\right)\right]. \tag{8.39}$$

However, $1 - |\gamma|^2 = 0$, and then

$$\phi_1^* \frac{d\phi_1}{dx} = ik_1|A|^2 \left(\gamma^* e^{2ik_1 x} - \gamma e^{-2ik_1 x}\right). \tag{8.40}$$

By taking the complex conjugate of the above equation, we obtain

$$\phi_1 \frac{d\phi_1^*}{dx} = -ik_1|A|^2 \left(\gamma e^{-2ik_1 x} - \gamma^* e^{2ik_1 x}\right). \tag{8.41}$$

Thus,

$$\phi_1^* \frac{d\phi_1}{dx} - \phi_1 \frac{d\phi_1^*}{dx} = 0, \tag{8.42}$$

and then

$$J_1 = \frac{\hbar}{2im}\left(\phi_1^* \frac{d\phi_1}{dx} - \phi_1 \frac{d\phi_1^*}{dx}\right) = 0. \tag{8.43}$$

In the region II, the wave function of the particles is

$$\phi_2(x) = \frac{2A}{(1 + i\beta)} e^{-k_2 x}. \tag{8.44}$$

Hence

$$\frac{d\phi_2}{dx} = \frac{-2k_2 A}{(1 + i\beta)} e^{-k_2 x}, \tag{8.45}$$

and then

$$\phi_2^* \frac{d\phi_2}{dx} = \frac{-4k_2|A|^2}{1+\beta^2} e^{-2k_2 x} . \tag{8.46}$$

Since $\phi_2^* \frac{d\phi_2}{dx}$ is a real function

$$\phi_2^* \frac{d\phi_2}{dx} = \phi_2 \frac{d\phi_2^*}{dx} , \tag{8.47}$$

and then

$$J_2 = \frac{\hbar}{2im} \left(\phi_2^* \frac{d\phi_2}{dx} - \phi_2 \frac{d\phi_2^*}{dx} \right) = 0 . \tag{8.48}$$

Thus, in both regions the probability current is zero.

The case: $E > V_0$.

We now turn to the case in which the particles have energy larger than the potential barrier. Since $E > V_0$, one could expect that the particles should travel freely from region I to region II. This is true in classical mechanics, but it is not true in quantum mechanics. We will show that a part of the particles can be reflected from the barrier.

First, we calculate the wave function of the particles. Since in region II the energy E of the particles is larger than the potential barrier, the parameter k^2 is a positive number and, therefore, the solutions to the Schrödinger equation in the two regions are of the form

$$\begin{aligned} \text{I.} \quad & \phi_1(x) = A e^{ik_1 x} + B e^{-ik_1 x} , & x < 0 \\ \text{II.} \quad & \phi_2(x) = C e^{ik_2 x} + D e^{-ik_2 x} , & x \geq 0 , \end{aligned} \tag{8.49}$$

where $k_1 = \sqrt{2mE}/\hbar$, and $k_2 = \sqrt{2m(E - V_0)}/\hbar$.

The particles, after passing to region II, will continue to move to the right with no reasons to turn back and move to the left, so we can put $D = 0$ in the wave function ϕ_2. Using the continuity conditions for the wave function, we obtain two equations for the coefficients $A, B,$ and C:

$$A + B = C ,$$
$$ik_1(A - B) = ik_2 C , \tag{8.50}$$

from which we find that

$$B = \frac{1-\beta}{1+\beta} A , \qquad C = \frac{2}{1+\beta} A , \tag{8.51}$$

where $\beta = k_2/k_1$.

Thus, for $E > V_0$, the wave function of the particles is given by

$$\phi_1(x) = A e^{ik_1 x} + \frac{(1 - \beta)^2}{1 + \beta} A e^{-ik_1 x}, \qquad x < 0 \qquad (8.52)$$

$$\phi_2(x) = \frac{2A}{(1 + \beta)} e^{ik_2 x}, \qquad x \geq 0 \qquad (8.53)$$

and, as usual, the coefficient A can be found from the normalization condition.

It is interesting to calculate the probability current in both regions, $x < 0$ and $x > 0$.

Consider first the probability current in region I:

$$J_1 = \frac{\hbar}{2im} \left(\phi_1^* \frac{d\phi_1}{dx} - \phi_1 \frac{d\phi_1^*}{dx} \right). \qquad (8.54)$$

Since

$$\frac{d\phi_1}{dx} = ik_1 A \left(e^{ik_1 x} - u e^{-ik_1 x} \right), \qquad (8.55)$$

where

$$u = \frac{1 - \beta}{1 + \beta}, \qquad (8.56)$$

we obtain

$$\phi_1^* \frac{d\phi_1}{dx} = ik_1 |A|^2 \left[(1 - u^2) - u e^{2ik_1 x} + u e^{-2ik_1 x} \right]. \qquad (8.57)$$

Hence

$$J_1 = \frac{\hbar |A|^2 k_1}{2m} 2(1 - u^2) = \frac{\hbar |A|^2 k_1}{m} (1 - u^2). \qquad (8.58)$$

In region II: $x \geq 0$, the wave function of the particles is

$$\phi_2(x) = \frac{2A}{(1 + \beta)} e^{ik_2 x}. \qquad (8.59)$$

Thus,

$$\frac{d\phi_2}{dx} = \frac{2ik_2 A}{1 + \beta} e^{ik_2 x}, \qquad (8.60)$$

and then

$$\phi_2^* \frac{d\phi_2}{dx} = \frac{4ik_2}{(1 + \beta)^2} |A|^2. \qquad (8.61)$$

Hence

$$J_2 = \frac{4\hbar k_2}{m(1 + \beta)^2} |A|^2. \qquad (8.62)$$

Since

$$1 - u^2 = \frac{4\beta}{(1+\beta)^2} , \tag{8.63}$$

we see that $J_1 = J_2$.

Note that the current in both regions is the same, as should be, otherwise particles would be created or destroyed.

We can find the transmission and reflection coefficients.

For $E < V_0$, it is obvious that the transmission coefficient $T = 0$, as the transmitted current is zero (see Eq. (8.48)).

The reflection coefficient is

$$R = \frac{|B|^2}{|A|^2} = \frac{(1 - i\beta)^2 (1 + i\beta)^2}{(1+\beta^2)^2} = 1 . \tag{8.64}$$

Thus, the relation $T + R = 1$ is satisfied.

For $E > V_0$, the transmission coefficient is

$$T = \frac{k_2 |C|^2}{k_1 |A|^2} = \frac{4k_2}{k_1 (1+\beta)^2} = \frac{4\beta}{(1+\beta)^2} , \tag{8.65}$$

and the reflection coefficient is

$$R = \frac{|B|^2}{|A|^2} = \frac{(1-\beta)^2}{(1+\beta)^2} . \tag{8.66}$$

Note that $T + R = 1$ should be always satisfied.

It is interesting that the transmission and reflection coefficients are independent of the mass of the particles.

To show this, we rewrite the transmission and reflection coefficients in terms of the parameters E, m, and V_0. Since

$$T = \frac{4\beta}{(1+\beta)^2} = \frac{4k_1 k_2}{(k_1 + k_2)^2} ,$$

$$R = \frac{(1-\beta)^2}{(1+\beta)^2} = \frac{(k_1 - k_2)^2}{(k_1 + k_2)^2} , \tag{8.67}$$

we substitute for k_1 and k_2

$$k_1 = \sqrt{\frac{2mE}{\hbar^2}} , \qquad k_2 = \sqrt{\frac{2m(E - V_0)}{\hbar^2}} , \tag{8.68}$$

we find

$$T = \frac{4\sqrt{E(E - V_0)}}{\left(\sqrt{E} + \sqrt{E - V_0} \right)^2} ,$$

$$R = \frac{\left(\sqrt{E} - \sqrt{E - V_0} \right)^2}{\left(\sqrt{E} + \sqrt{E - V_0} \right)^2} . \tag{8.69}$$

We see that the transmission and reflection coefficients are independent of the mass of the particles. They depend only on E and V_0. It is interesting—but someone can say very puzzling— that in some situations, like quantum tunneling (see Eq. (8.125) in Section 8.4), the coefficients depend on the mass of the particle, but in some situations, they do not depend on m.

An extra exercise

Just for fun, consider a similar problem as above, but now assume that the particles of energy $E > V_0$ are moving from $+x$ to $-x$, as shown in Fig. 8.5.

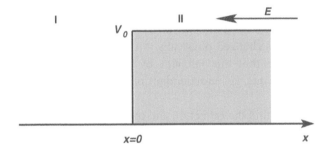

Figure 8.5 Potential step at $x = 0$. Particles of the total energy $E > V_0$ travel from $+x$ toward the potential step.

We expect that in the case of $E > V_0$, the behavior of the particles at the barrier should be the same, i.e., independent of the direction of motion of the particles. Is it true?

In order to check it, consider the solutions to the wave function in both regions. Since in both regions $E > V_0$, the wave functions in both regions are of the form of traveling waves

$$
\begin{array}{lll}
\text{I.} & \phi_2(x) = Ce^{ik_2x} + De^{-ik_2x}, & x \le 0, \\
\text{II.} & \phi_1(x) = Ae^{ik_1x} + Be^{-ik_1x}, & x > 0, \quad (8.70)
\end{array}
$$

where $k_1 = \sqrt{2m(E - V_0)}/\hbar$, and $k_2 = \sqrt{2mE}/\hbar$.

We can put $D = 0$, as we do not expect the particle to turn back and move to the right in region I.

Using the continuity conditions for the wave function, we obtain two equations for the coefficients A, B, and C:

$$A + B = C,$$

$$ik_1(A - B) = ik_2C, \qquad (8.71)$$

from which we find that

$$B = \frac{1 - \beta}{1 + \beta}A, \qquad C = \frac{2}{1 + \beta}A, \qquad (8.72)$$

where $\beta = k_2/k_1$. Thus, the wave function of the particles is given by

$$\phi_1(x) = Ae^{ik_1x} + \frac{(1 - \beta)^2}{1 + \beta}Ae^{-ik_1x}, \qquad x > 0 \qquad (8.73)$$

$$\phi_2(x) = \frac{2A}{(1 + \beta)}e^{ik_2x}, \qquad x \leq 0 \qquad (8.74)$$

and the coefficient A can be found from the normalization condition.

Note that the wave function (8.74) is the same as in Eq. (8.53). Thus, nothing is changed physically in this problem. This confirms our expectation that the behavior of the particle at the barrier should be the same, i.e., independent of the direction of motion of the particle.

We can also find the reflection and transmission coefficients and verify the relation $T + R = 1$.

Using the solutions to the coefficients C and B, we find that the transmission coefficient is

$$T = \frac{k_2|C|^2}{k_1|A|^2} = \frac{4k_2}{k_1(1 + \beta)^2} = \frac{4\beta}{(1 + \beta)^2}, \qquad (8.75)$$

and the reflection coefficient is

$$R = \frac{|B|^2}{|A|^2} = \frac{(1 - \beta)^2}{(1 + \beta)^2}, \qquad (8.76)$$

where $\beta = k_2/k_1$. Hence

$$T + R = \frac{4\beta}{(1 + \beta)^2} + \frac{(1 - \beta)^2}{(1 + \beta)^2} = \frac{(1 + \beta)^2}{(1 + \beta)^2} = 1. \qquad (8.77)$$

Thus, the relation $T + R = 1$ is satisfied.

Note that the transmission and reflection coefficients are the same we found before for the case of the particle moving from $-x$ to $+x$. Thus, again we can conclude that the behavior of the particles at the barrier is the same, i.e., independent of the direction of motion of the particles.

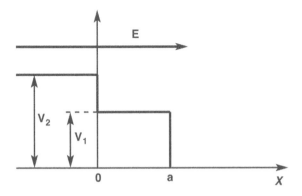

Figure 8.6   A double potential step.

Discussion Problem

Problem D3 We have already illustrated the phenomena of reflection and transmission of a group of particles acting on a potential barrier. One can notice that these phenomena are independent of the number of particles, so that the results may as well be applied to a single particle. Then, an obvious question arises: how to explain simultaneously that both the transmission and reflection coefficients of the particle are different from zero? Does it mean that a part of the particle is reflected and a part is transmitted?

Tutorial Problems

Problem 8.7 *Double potential step*
Particles of mass m and energy E moving in one dimension from $-x$ to $+x$ encounter a double potential step, as shown in Fig. 8.6, where

$$V_1 = \frac{\pi^2 \hbar^2}{8ma^2}, \qquad E = 2V_1, \qquad V_1 < V_2 < E.$$

(a) Find the transmission coefficient T.
(b) Find the value of V_2 at which T is maximal.

In the quantum world there is more than one mystery.

—H. P. Silverman

8.3 Finite Square-Well Potential

The infinite potential well, discussed in Section 8.1, is an idealized example. More realistic problems in physics have finite energy barriers, such as the step potential discussed in the preceding section, or more complicated arrangements composed of two or multiple potential steps. In such systems, one of the most interesting differences between classical and quantum descriptions of behavior of particles concerns the phenomenon of barrier penetration and transmission of particles through the barrier from one region to others.

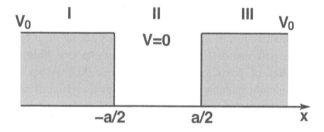

Figure 8.7   Finite square-well potential.

In this chapter, we will consider a particle moving in a finite square-well potential, as shown in Fig. 8.7:

$$
\begin{array}{lll}
\text{I.} & V(x) = V_0, & x < -\dfrac{a}{2} \\[2mm]
\text{II.} & V(x) = 0, & -\dfrac{a}{2} \le x \le \dfrac{a}{2} \\[2mm]
\text{III.} & V(x) = V_0, & x > \dfrac{a}{2}.
\end{array}
\qquad (8.78)
$$

In classical physics, a particle is trapped in the well if the energy E of the particle is less than V_0. In this case, the probability of finding the particle outside the well is zero. When the energy E is larger than V_0, the particle can freely move in all three regions. Let us look at these situations from the point of view of quantum physics.

Before proceeding to a rigorous calculation of the particle wave function, we point out that in behavior of the particle in a finite square-well potential, it must be recognized that the wave function of the particle exists in all space, that all regions in space are accessible for the particle even if the energy E is less than V_0. We have shown before that in the limit of $V_0 \to \infty$, the wave function is zero at the walls and in the region of infinite potential. In the present case, when the confining potential has a finite value, the particle wave function does not equal zero at the walls and can be different from zero in the regions where $E < V_0$.

To obtain a rigorous solution to the wave function, we will consider the Schrödinger equation for the wave function of the particle in all three regions indicated in Fig. 8.7.

In region II, $-a/2 \leq x \leq a/2$, the potential $V(x) = 0$, and then the Schrödinger equation reduces to

$$\frac{d^2\phi_2(x)}{dx^2} = -k_2^2\phi_2(x) , \tag{8.79}$$

where $k_2^2 = 2mE/\hbar^2$, and $\phi_2(x)$ is the wave function of the particle in region II.

Since k_2^2 is positive, the solution to Eq. (8.79) is of the form

$$\phi_2(x) = Ae^{ik_2x} + Be^{-ik_2x} , \tag{8.80}$$

that is, the same as for the particle in the potential well. Thus, we expect that, similar to the case of the infinite potential barrier, the energy of the particle will be quantized in region II.

In regions I and III, the potential is different from zero $V(x) = V_0$, and then

$$k_1^2 = -\frac{2m}{\hbar^2}(V_0 - E) = \frac{2m}{\hbar^2}(E - V_0) . \tag{8.81}$$

In this case, the Schrödinger equation is given by

$$\frac{d^2\phi_i(x)}{dx^2} = -k_1^2\phi_i(x) , \qquad i = 1, 3 . \tag{8.82}$$

Solution to the above equation depends on the relation between V_0 and E.

Consider separately two cases: $E > V_0$ and $E < V_0$.

8.3.1 *The case $E > V_0$*

For $E > V_0$, the parameter k_1^2 is positive, and then the solution to Eq. (8.82) in regions I and III is of the form

$$\text{I. } \phi_1(x) = Ce^{ik_1x} + De^{-ik_1x} ,$$
$$\text{III. } \phi_3(x) = Fe^{ik_1x} + Ge^{-ik_1x} , \tag{8.83}$$

indicating that the probability of finding the particle in regions I and III is similar to that in region II. It is not difficult to show that in this case the energy spectrum of the particle is continuous in all regions. Thus, one could conclude that there is nothing particularly interesting about the solution when $E > V_0$. However, we may obtain nonzero reflection coefficient at the boundaries, which is a quantum effect. Classically, one would expect that the particle of energy $E > V_0$ should travel from region I to region III without any reflection at the boundaries, and that the transmission coefficient should be unity, $T = 1$. It can be shown that only under a specific condition, the transmission coefficient becomes unity.[a]

8.3.2 *The case $E < V_0$*

More interesting is the case of $E < V_0$, i.e., when the particles have energy smaller than the potential barrier. When $E < V_0$, the parameter k_1^2 is a negative real number. In this case, the solution to Eq. (8.82) in regions I and III is in the form of nonoscillatory exponential functions

$$\phi_1(x) = Ce^{k_1x} + De^{-k_1x} , \qquad x < -\frac{a}{2}$$
$$\phi_3(x) = Fe^{k_1x} + Ge^{-k_1x} , \qquad x > \frac{a}{2} . \tag{8.84}$$

To get $\phi_1(x)$ and $\phi_3(x)$ finite for each $|x| > a/2$, in particular at $x \to \pm\infty$, we have to choose $D = F = 0$. Otherwise, the wave function would be infinite at $x = \pm\infty$. Hence, the acceptable solutions to the wave function in regions I and III are

$$\text{I. } \phi_1(x) = Ce^{k_1x} , \qquad \text{for} \quad x < -\frac{a}{2}$$
$$\text{III. } \phi_3(x) = Ge^{-k_1x} , \qquad \text{for} \quad x > \frac{a}{2} . \tag{8.85}$$

---

[a]The details of the calculations are left for the readers as a tutorial problem.

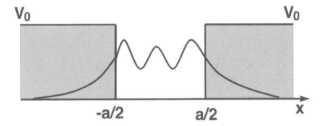

Figure 8.8 An example of the probability function of the particle inside the potential well.

Assume for a moment that C, $G \neq 0$. Then the probability density $|\phi(x)|^2$ has interesting properties, as shown in Fig. 8.8.

Referring to Fig. 8.8, the most striking features of the wave function $\phi(x)$ are the "tails" that extend outside the well. The nonzero values of $\phi(x)$ outside the well mean that there is a nonzero probability for finding the particle in regions I and III. In other words, the reflection of the particle takes place within the barriers, not at their surfaces.

Note that regions I and III are forbidden by classical physics because the particle would have to have negative kinetic energy. Since the total energy of the particle $E < V_0$ and $E = E_k + V_0$, we have

$$E_k + V_0 < V_0 \qquad \text{in the regions I and III}. \qquad (8.86)$$

Therefore, the penetration of the barrier is a quantum effect that has no classical analog. It is a strange effect, which makes a quantum mechanical penetration of the barrier to be regarded as a paradoxical, controversial, nonintuitive aspect of quantum physics.

How far the particle can penetrate the barrier?

This depends on V_0 and the mass of the particle. To show this, consider the wave function in region III. In this case,

$$\phi_3(x) = Ge^{-k_1 x}, \qquad x > \frac{a}{2}, \qquad (8.87)$$

with

$$k_1 = \frac{1}{\hbar}\sqrt{2m(V_0 - E)}. \qquad (8.88)$$

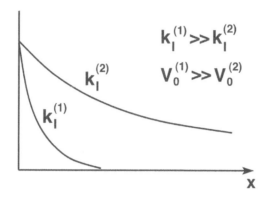

Figure 8.9 The dependence of $\exp(-k_1 x)$ on x for two different values of k_1.

Since the parameter k_1 is a positive real number, it plays a role of the damping coefficient of the exponential function. For $V_0 \gg E$, the parameter $k_1 \gg 1$, and then the depth of penetration is very small (vanishes for $V_0 \to \infty$). For $V_0 \approx E$, the parameter $k_1 \approx 0$, and then the depth of penetration is very large. These two situations are shown in Fig. 8.9.

The same arguments apply for the mass m of the particle. For a heavy particle, $m \gg 1$, the parameter $k_1 \gg 1$, and then the depth of penetration is very small.

To prove that the penetration effect really exists, we have to demonstrate that the constants C and G are really nonzero.

To show this, we will turn to the details and carry out the complete solution to the wave function of the particle.

We start from the general solution to the Schrödinger equation for the wave function of the particle inside the square-well potential, which is of the form:

$$\text{I.} \quad \phi_1(x) = C e^{k_1 x}, \qquad x < -\frac{a}{2}$$

$$\text{II.} \quad \phi_2(x) = A e^{ik_2 x} + B e^{-ik_2 x}, \qquad -\frac{a}{2} \le x \le \frac{a}{2}$$

$$\text{III.} \quad \phi_3(x) = G e^{-k_1 x}, \qquad x > \frac{a}{2}. \tag{8.89}$$

To find the constants A, B, C, and G, we use the property of the wave function that $\phi(x)$ and the first-order derivative $d\phi(x)/dx$

must be finite and continuous everywhere, in particular, at the boundaries $x = -a/2$ and $x = a/2$.

Hence, at $x = -a/2$:

$$Ce^{-\frac{1}{2}ak_1} = Ae^{-i\frac{1}{2}ak_2} + Be^{i\frac{1}{2}ak_2} . \qquad (8.90)$$

At $x = a/2$:

$$Ge^{-\frac{1}{2}ak_1} = Ae^{i\frac{1}{2}ak_2} + Be^{-i\frac{1}{2}ak_2} . \qquad (8.91)$$

We remember that also $d\phi/dx$ must be continuous across the same boundaries. Since

$$\text{I.} \quad \frac{d\phi_1}{dx} = k_1 C e^{k_1 x} ,$$

$$\text{II.} \quad \frac{d\phi_2}{dx} = ik_2 A e^{ik_2 x} - ik_2 B e^{-ik_2 x} ,$$

$$\text{III.} \quad \frac{d\phi_3}{dx} = -k_1 G e^{-k_1 x} , \qquad (8.92)$$

we find that at $x = -a/2$:

$$Ck_1 e^{-\frac{1}{2}ak_1} = ik_2 \left(Ae^{-\frac{1}{2}iak_2} - Be^{\frac{1}{2}iak_2} \right) , \qquad (8.93)$$

and at $x = a/2$:

$$-Gk_1 e^{-\frac{1}{2}ak_1} = ik_2 \left(Ae^{\frac{1}{2}iak_2} - Be^{-\frac{1}{2}iak_2} \right) . \qquad (8.94)$$

Dividing both sides of Eq. (8.93) by k_1, we obtain

$$Ce^{-\frac{1}{2}ak_1} = i\beta \left(Ae^{-\frac{1}{2}iak_2} - Be^{\frac{1}{2}iak_2} \right) , \qquad (8.95)$$

where $\beta = k_2/k_1$. Comparing Eqs. (8.90) and (8.95), we get

$$Ae^{-i\frac{1}{2}ak_2} + Be^{i\frac{1}{2}ak_2} = i\beta \left(Ae^{-\frac{1}{2}iak_2} - Be^{\frac{1}{2}iak_2} \right) , \qquad (8.96)$$

from which we find that

$$A = \frac{(i\beta + 1)}{(i\beta - 1)} Be^{iak_2} . \qquad (8.97)$$

With this relation between A and B, it is now possible to obtain a solution to the constant C in terms of B. Thus, substituting Eq. (8.97) into Eq. (8.90), we find that

$$C = \frac{2i\beta}{(i\beta - 1)} Be^{\frac{1}{2}ak_1(i\beta+1)} . \qquad (8.98)$$

Since $B \neq 0$, as the particle exists inside the well, we have $C \neq 0$, indicating that there is a nonzero probability of finding the particle in region I. The probability is given by $|\phi_1(x)|^2$, that is

$$|\phi_1(x)|^2 = |C|^2 e^{-2k_1|x|} , \tag{8.99}$$

where $|x| = -x$ for $x < 0$. Thus,

$$|\phi_1(x)|^2 = |B|^2 \frac{4\beta^2}{(\beta^2 + 1)} e^{-2k_1\left(|x| - \frac{a}{2}\right)} . \tag{8.100}$$

The probability is different from zero and decreases exponentially with the rate $2k_1$. The remaining constant $|B|$ is found from the normalization condition of $\phi(x)$.

We may conclude that even though the particle's energy is smaller than the value of V outside the well, there is still a definite probability that the particle can be found outside the well.

8.3.3 *Discrete Energy Levels*

We now check whether the constant G is different from zero and that the particle can really be found in region III. The calculations will also give us the condition for energies of the particle inside the well.

To find G, we consider the continuity conditions at $x = a/2$. From the symmetry of the system, we expect that the constant G, similar to C, will be different from zero and may be found in a similar way as we have found the constant C.

However, from Eqs. (8.91) and (8.94), and using Eq. (8.97), we find *two* solutions to the constant G in terms of B:

$$Ge^{-\frac{1}{2}ak_1} = B \left(ue^{\frac{3}{2}iak_2} + e^{-\frac{1}{2}iak_2} \right) , \tag{8.101}$$

$$Ge^{-\frac{1}{2}ak_1} = -i\beta B \left(ue^{\frac{3}{2}iak_2} - e^{-\frac{1}{2}iak_2} \right) , \tag{8.102}$$

where

$$u = \frac{(i\beta + 1)}{(i\beta - 1)} . \tag{8.103}$$

Thus, we have two different solutions to G. However, we cannot accept both the solutions as it would mean that there are two different probabilities of finding the particle at a point x inside region III. Therefore, we have to find under which circumstances these two solutions are equal.

Dividing Eq. (8.102) by Eq. (8.101), we obtain

$$\frac{i\beta \left(ue^{2iak_2} - 1\right)}{\left(u\, e^{2iak_2} + 1\right)} = -1 , \tag{8.104}$$

from which we find that the solutions (8.101) and (8.102) are equal when

$$\tan(ak_2) = \frac{2k_1k_2}{k_2^2 - k_1^2} . \tag{8.105}$$

To proceed further, we introduce dimensionless parameters

$$\epsilon = \frac{1}{2}ak_2 = \frac{1}{2}a\sqrt{\frac{2mE}{\hbar^2}} , \tag{8.106}$$

$$\sqrt{\xi^2 - \epsilon^2} = \frac{1}{2}ak_1 , \tag{8.107}$$

where

$$\xi^2 = \frac{ma^2V_0}{2\hbar^2} . \tag{8.108}$$

We see from the relation $\epsilon = \frac{1}{2}ak_2$ that determining ϵ, we can get the energy E of the particle inside the well. To show this, we rewrite Eq. (8.105) in terms of ϵ and ξ, and find

$$\epsilon \tan \epsilon = \sqrt{\xi^2 - \epsilon^2} . \tag{8.109}$$

This is a *transcendental equation*, which cannot be solved analytically, but it can be solved graphically as follows. Introducing the notation

$$p(\epsilon) = \epsilon \tan \epsilon ,$$
$$q(\epsilon) = \sqrt{\xi^2 - \epsilon^2} , \tag{8.110}$$

we find solutions to the equation $p(\epsilon) = q(\epsilon)$ by plotting separately $p(\epsilon)$ and $q(\epsilon)$. The functions $p(\epsilon)$ and $q(\epsilon)$ are shown in Fig. 8.10. The intersection points of the two curves give the solutions to the equation $p(\epsilon) = q(\epsilon)$. We see from the figure that the equation $p(\epsilon) = q(\epsilon)$ is satisfied only for *discrete* (finite) values of ϵ. Since the energy E is proportional to ϵ, see Eq. (8.106), we find that the energy of the particle is quantized in region II, i.e., the energy spectrum is discrete.

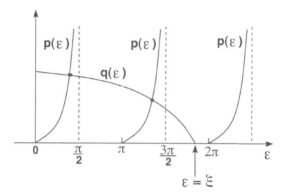

Figure 8.10 $p(\epsilon)$ and $q(\epsilon)$ as a function of ϵ. The number of solutions to the equation $p(\epsilon) = q(\epsilon)$ is given by the number of intersections of $p(\epsilon)$ and $q(\epsilon)$.

The number of solutions, which will give us the number of energy levels, depends on ξ, but always is finite. We see from Fig. 8.10 that we have

$$
\begin{array}{lll}
\text{one solution} & \text{for} & \xi < \pi, \\
\text{two solutions} & \text{for} & \xi < 2\pi, \\
\text{three solutions} & \text{for} & \xi < 3\pi,
\end{array}
$$

and so on. The number of solutions determines the number of energy levels inside the well. We remember that

$$
\xi^2 = \frac{1}{4}a^2 \left(k_1^2 + k_2^2 \right) = \frac{ma^2 V_0}{2\hbar^2}. \tag{8.111}
$$

Thus, ξ is determined by the dimensions of the well.

The number of energy levels that the well is capable of binding is often called the "strength" of the well. It is interesting to note from Fig. 8.10 that there must be at least one energy level inside the well and that the number of energy levels can never be zero.

Exercise in Class

Find the number of energy levels and the allowed values of E in a potential well with $\xi = 4$.

Since $\xi = 4 < 2\pi$, we see from Fig. 8.10 that in this case there are two solutions to ϵ: $\epsilon_1 = 1.25$ and $\epsilon_2 = 3.60$. Thus, there are only two energy levels inside the well.

Once we have the allowed values of ϵ, we can find the allowed values of the energy E. The calculations proceed in the following way. Since

$$\epsilon = \frac{1}{2}ak_2, \qquad \sqrt{\xi^2 - \epsilon^2} = \frac{1}{2}ak_1, \qquad (8.112)$$

and

$$k_1 = \sqrt{\frac{2m}{\hbar^2}(V_0 - E)}, \qquad k_2 = \sqrt{\frac{2m}{\hbar^2}E}, \qquad (8.113)$$

we get

$$\frac{\sqrt{\xi^2 - \epsilon^2}}{\epsilon} = \frac{k_1}{k_2} = \sqrt{\frac{V_0 - E}{E}}, \qquad (8.114)$$

from which we find that

$$E = \frac{\epsilon^2}{\xi^2}V_0. \qquad (8.115)$$

Hence, for $\epsilon_1 = 1.25$ and $\epsilon_2 = 3.60$, we get $E_1 = 0.098V_0$ and $E_2 = 0.81V_0$, respectively. The two energy levels are shown in Fig. 8.11.

Discussion Problems

Problem D4 Suppose someone would like to put a particle of energy E into the well analyzed in the aforementioned exercise in class.

What would happen to the particle of energy $E \neq E_1, E_2$?

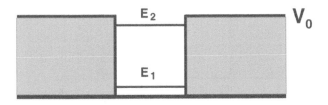

Figure 8.11 Energy levels inside the well with $\xi = 4$.

Problem D5 In the chapter, we have shown that the phenomenon of penetration of the barrier would imply that the kinetic energy of the particle in regions I and III is negative. As a consequence, the particle would have an imaginary speed.

Attack this problem from the standpoint of Heisenberg uncertainty principle and calculate the uncertainty in the kinetic energy of the particle in regions I and III. Compare the uncertainty in E_k to the amount of E_k; we expect it to be negative. How would you comment on this result?

Tutorial Problems

Problem 8.8 Show that the particle probability current density $\vec{J}$ is zero in region I, and deduce that $R = 1$, $T = 0$. This is the case of total reflection; the particle coming toward the barrier will eventually be found moving back. "Eventually," because the reversal of direction is not sudden. Quantum barriers are "spongy" in the sense the quantum particle may penetrate them in a way that classical particles may not.

Problem 8.9 Recall the case of $E > V_0$ discussed briefly in Section 8.3.1.

(a) Evaluate the transmission coefficient from region I to region III.

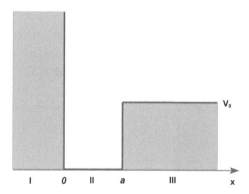

Figure 8.12 Potential well of semi-infinite depth.

(b) Under which condition the transmission coefficient becomes unity, $T = 1$?

Problem 8.10 A rectangular potential well is bounded by a wall of infinite high on one side and a wall of high V_0 on the other, as shown in Fig. 8.12. The well has a width a and a particle located inside the well has energy $E < V_0$.

(a) Find the wave function of the particle inside the well.
(b) Show that the energy of the particle is quantized.
(c) Discuss the dependence of the number of energy levels inside the well on V_0.

> *Quantum mechanics is magic.*
> —Daniel Greenberger

8.4 Quantum Tunneling

When the potential barrier has a finite width, an interesting effect can appear, that the particle with energy $E < V_0$ can not only penetrate the barrier, it can even pass through the barrier and appear on the other side, as illustrated in Fig. 8.13.

This phenomenon is known as *quantum tunneling*. In solid state physics, this phenomenon is often called *cold emission*.

How can we understand the process of a particle tunneling through a seemingly impenetrable barrier? How large is the probability that the particle passes through the barrier? To answer these questions, we can use a wave picture of the particle. We have already learned that the particle wave function does not terminate abruptly at the edge of the barrier, but actually leaks out of the barrier. As we will see, the tunneling effect depends on the relation between E and V_0, mass m of the particle, and also on the thickness of the barrier.

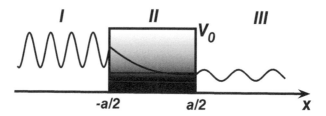

Figure 8.13 Potential barrier of a thickness a. The wave function of a particle is continuous across the barrier, which may result the particle to tunnel through the barrier from region I to region III.

Rigorous calculations

The analytic formulation of the tunneling effect can be obtained simply as follows: Consider a particle of mass m and total energy E moving from the left and acting on the rectangular potential barrier, as shown in Fig. 8.13. Assume that $E < V_0$. According to classical physics, the particle cannot pass the barrier. What does the quantum mechanics say about it?

According to quantum mechanics, the particle is determined by the wave function, which exists in all three regions shown in Fig. 8.13. The wave function is found from the Schrödinger equation, whose solution for the three regions is

$$\text{I.} \quad \phi_1(x) = Ae^{ik_1x} + Be^{-ik_1x}, \qquad x < -\frac{a}{2}$$

$$\text{II.} \quad \phi_2(x) = Ce^{-k_2x} + De^{k_2x}, \qquad -\frac{a}{2} \leq x \leq \frac{a}{2}$$

$$\text{III.} \quad \phi_3(x) = Fe^{ik_1x} + Ge^{-ik_1x}, \qquad x > \frac{a}{2} \qquad (8.116)$$

where $k_1 = \sqrt{2mE}/\hbar$ and $k_2 = \sqrt{2m(V_0 - E)}/\hbar$.

The wave function and its first-order derivatives must be continuous at the two boundaries $x = -a/2$ and $x = a/2$. The continuity conditions at $x = -a/2$ are

$$Ae^{-ik_1\frac{a}{2}} + Be^{ik_1\frac{a}{2}} = Ce^{k_2\frac{a}{2}} + De^{-k_2\frac{a}{2}},$$

$$Ae^{-ik_1\frac{a}{2}} - Be^{ik_1\frac{a}{2}} = \frac{ik_2}{k_1}\left(Ce^{k_2\frac{a}{2}} - De^{-k_2\frac{a}{2}}\right). \qquad (8.117)$$

The continuity conditions at $x = a/2$ are

$$Ce^{-k_2\frac{a}{2}} + De^{k_2\frac{a}{2}} = Fe^{ik_1\frac{a}{2}} + Ge^{-ik_1\frac{a}{2}},$$

$$-Ce^{-k_2\frac{a}{2}} + De^{k_2\frac{a}{2}} = \frac{ik_1}{k_2}\left(Fe^{ik_1\frac{a}{2}} - Ge^{-ik_1\frac{a}{2}}\right). \quad (8.118)$$

The continuity conditions (8.117) and (8.118) can be concisely written in terms of matrix equations that are easy to solve:

$$\begin{pmatrix} A \\ B \end{pmatrix} = \frac{1}{2}\begin{pmatrix} \left(1 + \frac{ik_2}{k_1}\right)e^{k_2\frac{a}{2}+ik_1\frac{a}{2}} & \left(1 - \frac{ik_2}{k_1}\right)e^{-k_2\frac{a}{2}+ik_1\frac{a}{2}} \\ \left(1 - \frac{ik_2}{k_1}\right)e^{k_2\frac{a}{2}-ik_1\frac{a}{2}} & \left(1 + \frac{ik_2}{k_1}\right)e^{-k_2\frac{a}{2}-ik_1\frac{a}{2}} \end{pmatrix}\begin{pmatrix} C \\ D \end{pmatrix},$$

$$(8.119)$$

and

$$\begin{pmatrix} C \\ D \end{pmatrix} = \frac{1}{2}\begin{pmatrix} \left(1 - \frac{ik_1}{k_2}\right)e^{k_2\frac{a}{2}+ik_1\frac{a}{2}} & \left(1 + \frac{ik_1}{k_2}\right)e^{k_2\frac{a}{2}-ik_1\frac{a}{2}} \\ \left(1 + \frac{ik_1}{k_2}\right)e^{-k_2\frac{a}{2}+ik_1\frac{a}{2}} & \left(1 - \frac{ik_1}{k_2}\right)e^{-k_2\frac{a}{2}-ik_1\frac{a}{2}} \end{pmatrix}\begin{pmatrix} F \\ G \end{pmatrix}.$$

$$(8.120)$$

The relationship between the solution on the left of the barrier and the solution on the right can now be obtained by substituting the matrix equation (8.120) into the right-hand side of the matrix equation (8.119). This leads to the matrix equation

$$\begin{pmatrix} A \\ B \end{pmatrix} = \begin{pmatrix} M_{11} & M_{12} \\ M_{12}^* & M_{11}^* \end{pmatrix}\begin{pmatrix} F \\ G \end{pmatrix}, \quad (8.121)$$

where

$$M_{11} = \left(\cosh k_2 a + \frac{i\epsilon}{2}\sinh k_2 a\right)e^{ik_1 a},$$

$$M_{12} = \frac{i\eta}{2}\sinh k_2 a, \quad (8.122)$$

with

$$\epsilon = \frac{k_2}{k_1} - \frac{k_1}{k_2}, \quad \text{and} \quad \eta = \frac{k_2}{k_1} + \frac{k_1}{k_2}. \quad (8.123)$$

We now have four unknown coefficients to evaluate. But we have only two equations and a third from the normalization. However, if we assume that particles travel from the left to the right, we can set $G = 0$ as there is nothing in region III that could reflect particles back toward the barrier. In this case, it is a simple algebraic exercise to show that

$$\frac{F}{A} = \frac{e^{-ik_1 a}}{\cosh(k_2 a) + \frac{1}{2} i \epsilon \sinh(k_2 a)}. \tag{8.124}$$

Hence, the transmission coefficient of the particle from region I to region III is

$$T = \frac{|F|^2}{|A|^2} = \left[\cosh^2(k_2 a) + \left(\frac{\epsilon}{2} \right)^2 \sinh^2(k_2 a) \right]^{-1}$$

$$= \left[1 + \left(\left(\frac{\epsilon}{2} \right)^2 + 1 \right) \sinh^2(k_2 a) \right]^{-1}$$

$$= \left[1 + \frac{V_0^2 \sinh^2(k_2 a)}{4E(V_0 - E)} \right]^{-1}. \tag{8.125}$$

When $E > V_0$, the solution changes only in region II. In this case, the propagation constant k_2 is imaginary, and the new solution can be readily obtained from the above solution simply by replacing k_2 by ik_2', where $k_2' = \sqrt{2m(E - V_0)}/\hbar$. So that the expression for T changes to

$$T = \left[1 + \frac{V_0^2 \sin^2(k_2' a)}{4E(E - V_0)} \right]^{-1}. \tag{8.126}$$

First, note that if there is no barrier, $V_0 = 0$, and then there is 100% transmission. Thus, there is nothing particularly remarkable about the solution when $V_0 = 0$. Its physical interest lies in what happens when $V_0 \neq 0$.

For $V_0 \neq 0$, one could expect from the classical physics that there is no transmission through the barrier, $T = 0$ for $E < V_0$, and that the transmission coefficient $T = 1$ for $E > V_0$. However, Eq. (8.125) shows that T is not zero for $E < V_0$ (nonzero transmission), and Eq. (8.126) shows that $T < 1$ for $E > V_0$ (nonzero reflection).

Thus, tunneling effect for $E < V_0$ and partial reflection at the barrier for $E > V_0$ are quantum phenomena. In the classical limit of $\hbar \to 0$, the parameter $k_2 \to \infty$, and then $T \to 0$. Hence, in the classical limit, there is no possibility of the particle of an energy $E < V_0$ to pass the barrier. Moreover, since that k_2 is proportional to the mass of the particle, the transmission coefficient (8.125) is large for light particles and decreases with the increasing m.

There is a further interesting phenomenon for the case of $E > V_0$. One can see from Eq. (8.126) that if $k_2'a = n\pi$, $n = 1, 2, 3, \ldots$, the transmission coefficient becomes unity, $T = 1$, i.e., there is 100% transmission from region I to region III. Otherwise, quantum effect (partial reflection) appears. The conditions at which there is 100% transmission are called *resonances*. This phenomenon is analogous to the behavior of coated lenses in optics. Another example is the scattering of electrons by atoms, where it is observed that for certain energies of the electrons, almost no scattering takes place, i.e., the atom is almost transparent. These are real examples of the transmission resonances, and in electron scattering by electrons, it is known as the *Townsend–Ramsauer effect*.

Remember: External potential (energy supply) is not required to tunnel particles through the barrier. Quantum tunneling does not use energy!

8.4.1 *Applications of Quantum Tunneling*

Quantum tunneling is important in the understanding of a number of physical phenomena such as thermonuclear reactions and conduction in metals and semiconductors. In 1928, Gamov, Condon, and Gurney used quantum tunneling to explain the α-decay of unstable nuclei, in which an alpha particle (a helium nucleus) is emitted from a radioactive nucleus. The alpha particle has an energy that is less than the high of the potential barrier keeping the particle inside the nucleus. Nevertheless, alpha particles tunnel through the barrier and are detected outside the nucleus.

Tunneling has also been used in a number of electronic devices. One is a tunnel diode in which the current of electrons is controlled by adjusting the energy E of the electrons. This changes the value of k_2 and thus the rate at which the electrons tunnel through the device.

The most advanced application of quantum tunneling is the scanning tunneling microscope. A probe needle is held very close (<1 nm) above a conducting object and scanned across it. The object is at a positive voltage V with respect to the probe needle. However, for electrons to pass from the needle to the object, they have to overcome the work function of the needle material. This creates a potential barrier through which electrons can tunnel. As they tunnel through the potential barrier, they generate a current whose variation tells us about the distance between the needle and the object.

Discussion Problem

Problem D6 Take the limits $V_0 \to \infty$ and $a \to 0$ in the solution (8.125) such that aV_0 is equal to a constant α. How does the transmission coefficient depend on E?

This problem illustrates transmission through a potential barrier of the shape of a δ function.

Tutorial Problems

Problem 8.11 Check the derivation of Eq. (8.121) from Eqs. (8.119) and (8.120).

Problem 8.12 Fill the missing steps in the derivation of Eq. (8.124) from Eq. (8.121).

Problem 8.13 Using either Eq. (8.125) or Eq. (8.126), find the transmission coefficient in the limit of $E \to V_0$.

Problem 8.14 An electron, which has a kinetic energy of 100 eV, enters a region of width $a = 0.1$ nm, where there is an accumulation of negative charge. The region may be treated as a potential barrier of 120 eV. Find the probability that the electron will be found on the other side of the barrier.

Problem 8.15 *Tunneling through a nonsymmetric barrier*
Particles of mass m and energy $E < V_0$ moving in one dimension from $-x$ to $+x$ encounter a nonsymmetric barrier, as shown in Fig. 8.14.

(a) Find the transmission coefficient T.
(b) Show that in the limit of $a \to 0$, the transmission coefficient reduces to that of the step potential.
(c) Does the transmission coefficient depend on the direction of propagation of the particles?

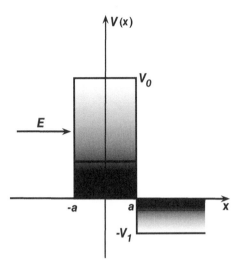

Figure 8.14 Tunneling through a nonsymmetric barrier.

Challenging problem: *Quantum tunneling from and into a semi-finite well*

In the problem discussed in the chapter, we have shown there are no restrictions on the energy of a particle to tunnel through the barrier. The explanation of this effect is simple: The particle of an arbitrary energy can tunnel through the barrier because there are no restrictions on energies, which the particle can have in region III. Let us consider the exercise from the preceding section—a rectangular potential well of width a bounded by a wall of infinite high on one side and a barrier of high V_0 and infinite thickness on the other. We have learned that inside the well in region I, a particle can have only a limited number of discrete energies.

(a) Now imagine what happens if the thickness of the barrier (region II) is finite and the particle of energy $E < V_0$ is inside the potential well. Do you expect that the energy levels of the particle in region I are still discrete?

(b) Suppose that the particle of energy $E < V_0$ is in region III on the right of the barrier and moves toward the barrier, as shown in Fig. 8.15.

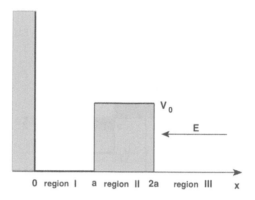

Figure 8.15   Potential well of semi-infinite depth.

Find the tunneling coefficient and determine whether for each E, in the range $0 < E < V_0$, the particle can tunnel to region I, i.e., determine whether the energy spectrum in region I is discrete or continuous. Would you expect this result without the detailed calculations?

Chapter 9

Multidimensional Quantum Wells

In the last three chapters, we have learned about some strange effects that occur in the one-dimensional quantum world. We have seen that particles confined into a small region can have quantized energies, can be found in "classically forbidden" region, and even can tunnel through this region. Although the one-dimensional case is very useful in the illustration and understanding of the quantum effects, we need a full three-dimensional treatment if we want to illustrate applications of quantum mechanics to atoms, solid state, nanotechnology, and nuclear physics. The application to atoms is postponed for last because it is more difficult and will be discussed in a separate chapter on angular momentum and hydrogen atom (Chapter 16).

In this chapter, we extend the concept of quantum wells from one to three dimensions to see how particles behave in a three-dimensional quantum world. We will consider properties of particles confined in two- and three-dimensional potential structures called quantum wires and quantum dots. Determining energies and corresponding wave functions requires some careful analysis but is worth the trouble.

We can picture a quantum wire as a pipe, as shown in Fig. 9.1, and particles moving freely along this pipe, just like water flowing

Quantum Physics for Beginners
Zbigniew Ficek
Copyright ⓒ 2016 Pan Stanford Publishing Pte. Ltd.
ISBN 978-981-4669-38-2 (Hardcover), 978-981-4669-39-9 (eBook)
www.panstanford.com

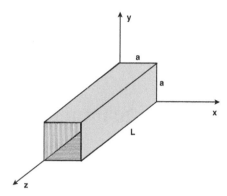

Figure 9.1   A three-dimensional well of sides $x = a$, $y = a$, and $z = L$. Inside the well, $V = 0$. The potential is infinite at the xy walls and can be set zero at the z walls (quantum wire), or infinite (quantum dot).

through a pipe. However, we must be careful when using analogies to describe quantum phenomena. We expect water to flow along the bottom of the pipe, but we would be surprised to see the water flowing through mid-air a few centimeters above the pipe floor. This is precisely how the particles appear to behave in a quantum wire.

9.1 General Solution to the Three-Dimensional Schrödinger Equation

Let us find the stationary wave function of a particle located inside a three-dimensional well and its energies. The readers will immediately notice that the wave function and the energies are found from the stationary three-dimensional Schrödinger equation

$$-\frac{\hbar^2}{2m}\nabla^2\Psi + V\Psi = E\Psi. \tag{9.1}$$

We see that in Cartesian coordinates, the operator $\partial^2/\partial x^2$ in the one-dimensional case is replaced in three dimensions by the Laplacian

$$\nabla^2 = \frac{\partial^2}{\partial x^2} + \frac{\partial^2}{\partial y^2} + \frac{\partial^2}{\partial z^2}. \tag{9.2}$$

Since x, y, z are independent (separable) variables, the wave function is also separable into three independent functions Ψ_x, Ψ_y, and Ψ_z. In this case, we can find the solution to the Schrödinger equation in product form

$$\Psi(x, y, z) = \Psi_x(x)\Psi_y(y)\Psi_z(z) . \tag{9.3}$$

Substituting this into the above Schrödinger equation and dividing both sides by $\Psi_x\Psi_y\Psi_z$, we obtain

$$-\frac{\hbar^2}{2m\Psi_x}\frac{d^2\Psi_x}{dx^2} - \frac{\hbar^2}{2m\Psi_y}\frac{d^2\Psi_y}{dy^2} - \frac{\hbar^2}{2m\Psi_z}\frac{d^2\Psi_z}{dz^2} = E , \tag{9.4}$$

where, as before, we have put $V = 0$ inside the well, but outside the well we will set $V = \infty$ as a boundary condition.

Since each term on the left-hand side of Eq. (9.4) is a function of only one variable, each will be independent of any change in the other two variables. Thus, Eq. (9.4) can be separated into three independent equations. To illustrate this, we write this equation as

$$-\frac{\hbar^2}{2m\Psi_x}\frac{d^2\Psi_x}{dx^2} = E + \frac{\hbar^2}{2m\Psi_y}\frac{d^2\Psi_y}{dy^2} + \frac{\hbar^2}{2m\Psi_z}\frac{d^2\Psi_z}{dz^2} . \tag{9.5}$$

The left-hand side involves all of the x dependence. If we change x any way we want, the right-hand side is not affected. Thus, it must be that both sides are equal to a constant, say E_x:

$$-\frac{\hbar^2}{2m\Psi_x}\frac{d^2\Psi_x}{dx^2} = E_x , \tag{9.6}$$

$$-\frac{\hbar^2}{2m\Psi_y}\frac{d^2\Psi_y}{dy^2} - \frac{\hbar^2}{2m\Psi_z}\frac{d^2\Psi_z}{dz^2} = E - E_x . \tag{9.7}$$

Equation (9.7), which depends only on y and z variables, can be written as

$$-\frac{\hbar^2}{2m\Psi_y}\frac{d^2\Psi_y}{dy^2} = E - E_x + \frac{\hbar^2}{2m\Psi_z}\frac{d^2\Psi_z}{dz^2} . \tag{9.8}$$

Again, both sides depend on different variables. The left-hand side depends only on y and the right-hand side depends only on z; thus, both sides are equal to a constant, say E_y:

$$-\frac{\hbar^2}{2m\Psi_y}\frac{d^2\Psi_y}{dy^2} = E_y , \tag{9.9}$$

$$-\frac{\hbar^2}{2m\Psi_z}\frac{d^2\Psi_z}{dz^2} = E - E_x - E_y . \tag{9.10}$$

Hence, after the separation of the variables, the complicated differential equation involving three variables has turned into three independent equations of one variable each:

$$-\frac{\hbar^2}{2m\Psi_x}\frac{d^2\Psi_x}{dx^2} = E_x \,, \tag{9.11}$$

$$-\frac{\hbar^2}{2m\Psi_y}\frac{d^2\Psi_y}{dy^2} = E_y \,, \tag{9.12}$$

$$-\frac{\hbar^2}{2m\Psi_z}\frac{d^2\Psi_z}{dz^2} = E_z \,, \tag{9.13}$$

where $E_z = E - E_x - E_y$.

The wave function of the particle inside the well and its energies are found from these three independent equations. The parameters E_x, E_y, and E_z are separation constants and represent energies of motion along the three Cartesian axes x, y, and z. It is easy to see that these three constants satisfy the equation

$$E_x + E_y + E_z = E \,. \tag{9.14}$$

The solutions to Eqs. (9.11) and (9.12) are the same as that for the infinite square-well in one dimension and are given by

$$\Psi_x = A \sin\left(\frac{n_1\pi}{a}x\right) \,, \qquad n_1 = 1, 2, 3, \dots \,, \tag{9.15}$$

$$\Psi_y = B \sin\left(\frac{n_2\pi}{a}y\right) \,, \qquad n_2 = 1, 2, 3, \dots \,, \tag{9.16}$$

with energies

$$E_x = n_1^2\frac{\pi^2\hbar^2}{2ma^2} \qquad \text{and} \qquad E_y = n_2^2\frac{\pi^2\hbar^2}{2ma^2} \,. \tag{9.17}$$

Thus, the energy of the particle is quantized in the x- and y-directions.

It is worth noting how inevitably quantum numbers appear in quantum theories of particles confined (trapped) into a particular region of space.

9.2 Quantum Wire

The solution to the z-component of the motion, Eq. (9.13), depends on whether the z sides of the well have zero or infinite potential. For zero potential, the z-direction is free for the motion of the particle. In this way we form a quantum wire. The z component of the wave function is given by

$$\Psi_z = C e^{ik_z z},$$ (9.18)

where $k_z^2 = 2mE_z/\hbar^2$ and E_z can have arbitrary values.

Thus, we conclude that for a quantum wire, the three-dimensional wave functions of the particle are of the form

$$\Psi_{n_1, n_2} = D \sin\left(\frac{n_1 \pi}{a} x\right) \sin\left(\frac{n_2 \pi}{a} y\right) e^{ik_z z},$$ (9.19)

and the corresponding energies are given by

$$E = \frac{\pi^2 \hbar^2}{2ma^2} \left(n_1^2 + n_2^2\right) + E_z, \qquad \text{with} \qquad n_1, n_2 = 1, 2, 3, \ldots,$$ (9.20)

where $D = ABC$ is a constant, which is found from the normalization condition of the wave function, and E_z can be arbitrary. It is easy to show that

$$D = \frac{2}{a}\sqrt{\frac{1}{L}}.$$ (9.21)

The proof of this formula is left as a tutorial problem for the readers.

Interesting observation: Since the energy of the particle in the y-direction can never be zero, the particle will never move at the floor of the wire. Because the particle is confined in two directions, it has only one dimension of freedom. Therefore, a quantum wire is sometimes referred to as a *one-dimensional* system.

9.3 Quantum Dots

Can we reduce the freedom of the particle still further? Yes, if the potential at the z sides is infinite, the particle is confined in every direction, so we have an example of quantum dot.[a] In this case, the solution to Eq. (9.13) is the same as for the x and y components:

$$\Psi_z = C \sin\left(\frac{n_3 \pi}{L} z\right), \qquad n_3 = 1, 2, 3, \ldots,$$ (9.22)

---

[a]It would be more correct to call the quantum dot a quantum box, but in many calculations, quantum dots that have spherical symmetries are approximated by rectangular boxes.

with the E_z energy taking only discrete values

$$E_z = n_3^2 \frac{\pi^2 \hbar^2}{2mL^2} \, . \tag{9.23}$$

Thus, for a quantum dot with $L = a$, the wave functions of the particle inside the well are of the form

$$\Psi_{n_1,n_2,n_3} = \left(\frac{2}{a}\right)^{\frac{3}{2}} \sin\left(\frac{n_1 \pi}{a}x\right) \sin\left(\frac{n_2 \pi}{a}y\right) \sin\left(\frac{n_3 \pi}{a}z\right) , \tag{9.24}$$

and the corresponding energies are given by

$$E = \frac{\pi^2 \hbar^2}{2ma^2} \left(n_1^2 + n_2^2 + n_3^2\right) , \qquad \text{with} \qquad n_1, n_2, n_3 = 1, 2, 3, \dots . \tag{9.25}$$

Notice that the energy of the particle in a quantum dot is quantized in all three directions. Because the motion of the particle is now restricted in every direction (quantum confinement), the particle has zero dimension of freedom. That is why a quantum dot is often referred to as a *zero-dimensional* system. Quantum dots are also regarded as *artificial atoms*.

An obvious question arises: Why are we interested in and what do we achieve by producing such a structure?

Currently, most physicists say that nothing exciting can be done with a single dot and it could have a little practical use. One can just observe discrete energies of the confined electron. This would only be another confirmation of the quantization of the energy. However, if a single dot is located on a chip, a set of closely located dots, it could have many practical uses. The set of quantum dots, composed into a regular array, can behave as a single macroscopic device performing a highly complex function. An electron located in one of the dots can tunnel to another dot and this process can be controlled by varying the energy levels in some of the dots. Thus, we can affect and control the flow of the current through the device just by changing energies of the dots. This is an example of an entirely different class of devices in which the current flow is due to the wave behavior of electrons.

9.3.1 Degeneracy of Energy Levels

One can notice that the results obtained for the three-dimensional case are similar to that obtained for the one-dimensional case.

However, there is a significant difference between these two cases. Notice from Eqs. (9.24) and (9.25) that in the three-dimensional case, the wave function and the energy are specified by the trio of integers (n_1, n_2, n_3). Thus, there might be few wave functions corresponding to the same energy. If it happens, we say that the energy level is **degenerate**, and the number of degenerate states, **degeneracy**.

We can illustrate this on a simple example of energies and the corresponding wave functions of a quantum dot:

The lowest energy state (the ground state), for which $n_1 = n_2 = n_3 = 1$, has energy

$$E = \frac{3\pi^2 \hbar^2}{2ma^2}, \qquad (9.26)$$

and there is only one wave function (singlet) corresponding to this energy. We say that the level has degeneracy 1.

However, for the first excited energy level, there are three wave functions corresponding to energy

$$E = \frac{6\pi^2 \hbar^2}{2ma^2}, \qquad (9.27)$$

as there are three combinations of n_1, n_2, and n_3 whose squares sum to 6. These combinations are $(n_1 = 2, n_2 = 1, n_3 = 1)$, $(n_1 = 1, n_2 = 2, n_3 = 1)$, and $(n_1 = 1, n_2 = 1, n_3 = 2)$. The three wave functions corresponding to this energy and specified by the three combinations are $\Psi_{2,1,1}$, $\Psi_{1,2,1}$, and $\Psi_{1,1,2}$. In this case, we say that the level has degeneracy 3.

An interesting observation: It is easy to notice from Eq. (9.25) that the degeneracy of the energy levels is characteristic of quantum wells whose sides are of equal lengths. The degeneracy can be lifted, if the sides of the well were of unequal lengths.

Revision Question

Quantum physics predicts that a particle can behave in unusual ways, which are regarded as nonclassical or quantum effects without classical analog.

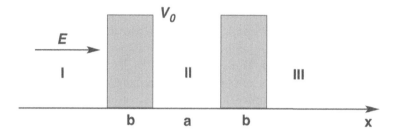

Figure 9.2   Tunneling through a quantum well.

(a) Give examples of nonclassical effects you have learned in the book.
(b) Explain using physical arguments, why these effects are regarded as nonclassical.

Tutorial Problems

Problem 9.1 This problem illustrates why the tunneling (flow) of an electron between different quantum dots is possible only for specific (discrete) energies of the electron.

Consider a simplified situation, a one-dimensional system of quantum wells, as shown in Fig. 9.2. The well represents a quantum dot. Show, using the method we learned in the previous chapters on applications of the Schrödinger equation, that an electron of energy $E < V_0$ and being in region I can tunnel through the quantum well (region II) to region III only if E is equal to the energy of one of the discrete energy levels inside the well.

Problem 9.2 Find the number of wave functions (energy states) of a particle in a quantum well of the sides of equal lengths corresponding to energy

$$E = \frac{9\pi^2 \hbar^2}{2ma^2},$$

i.e., for the combination of n_1, n_2, and n_3, whose squares sum to 9.

Problem 9.3 Find all energy states of a particle confined inside a three-dimensional box with energies below

$$15\frac{\pi^2\hbar^2}{2ma^2}.$$

Indicate the degeneracy of each energy level.

Chapter 10

Linear Operators and Their Algebra

We leave the stationary Schrödinger equation for a few chapters to learn some operator techniques and state representations commonly used in quantum physics. We have seen that in quantum physics, it is frequently required to calculate energy and momentum of a particle, which are represented by energy and momentum operators, respectively. In this chapter, we will extend the idea of representing physical as well as nonphysical quantities by operators and collect together some frequently used operator techniques. We will postulate that any quantity in quantum physics is specified by a linear operator.

An operator $\hat{A}$ is **linear** if for arbitrary functions f_i and g_i, and arbitrary complex numbers c_i, such that

$$\hat{A} f_1 = g_1 \qquad \text{and} \qquad \hat{A} f_2 = g_2 ,$$

the linear superposition is

$$\hat{A} (c_1 f_1 + c_2 f_2) = c_1 \hat{A} f_1 + c_2 \hat{A} f_2 = c_1 g_1 + c_2 g_2 . \qquad (10.1)$$

Quantum Physics for Beginners
Zbigniew Ficek
Copyright © 2016 Pan Stanford Publishing Pte. Ltd.
ISBN 978-981-4669-38-2 (Hardcover), 978-981-4669-39-9 (eBook)
www.panstanford.com

10.1 Algebra of Operators

Let $\hat{A}$ and $\hat{B}$ are two linear operators. The sum and product of the two operators $\hat{A}$ and $\hat{B}$ are defined as

$$\left(\hat{A} + \hat{B}\right) f = \hat{A}f + \hat{B}f, \qquad (10.2)$$
$$\left(\hat{A}\hat{B}\right) f = \hat{A}\left(\hat{B}f\right). \qquad (10.3)$$

The operators obey the following algebraic rules:

1. $\hat{A} + \hat{B} = \hat{B} + \hat{A}$,   (addition commutes),
2. $\hat{A} + \hat{B} + \hat{C} = \left(\hat{A} + \hat{B}\right) + \hat{C} = \hat{A} + \left(\hat{B} + \hat{C}\right)$,
3. $\hat{A}\hat{B}\hat{C} = \hat{A}\left(\hat{B}\hat{C}\right) = \left(\hat{A}\hat{B}\right)\hat{C}$, (associative law of multiplicity),
4. $\left(\hat{A} + \hat{B}\right)\hat{C} = \hat{A}\hat{C} + \hat{B}\hat{C}$,   (distributive law).

$$(10.4)$$

The power of an operator and the sum of two operators are defined as

1. $\hat{A}^2 = \hat{A}\hat{A}$, $\hat{A}^3 = \hat{A}\hat{A}\hat{A}$,   etc.
2. $\left(\hat{A} + \hat{B}\right)^2 = \left(\hat{A} + \hat{B}\right)\left(\hat{A} + \hat{B}\right) = \hat{A}^2 + \hat{B}^2 + \hat{A}\hat{B} + \hat{B}\hat{A}$.

$$(10.5)$$

In a manipulation with operators, it is very important to maintain the order of the operators because operator algebra is not in general commutative for multiplication. For example, $\hat{A}\hat{B} \neq \hat{B}\hat{A}$ in general and then

$$\left(\hat{A} + \hat{B}\right)^2 \neq \hat{A}^2 + \hat{B}^2 + 2\hat{A}\hat{B}, \qquad (10.6)$$

as in standard mathematics.

An important concept in quantum physics is a **commutator**. For two operators $\hat{A}$ and $\hat{B}$, the commutator is defined as

$$\left[\hat{A}, \hat{B}\right] = \hat{A}\hat{B} - \hat{B}\hat{A}, \qquad (10.7)$$

and says that two operators commute if

$$\left[\hat{A}, \hat{B}\right] = 0. \qquad (10.8)$$

We can also define *anti-commutator* as

$$\left\{\hat{A}, \hat{B}\right\} \equiv \left[\hat{A}, \hat{B}\right]_+ = \hat{A}\hat{B} + \hat{B}\hat{A}, \qquad (10.9)$$

and we say that two operators anti-commute if

$$\left[\hat{A}, \hat{B}\right]_+ = 0. \qquad (10.10)$$

Inverse operator: The inverse of an operator $\hat{A}$, if it exists, is defined by

$$\hat{A}\hat{A}^{-1} = \hat{A}^{-1}\hat{A} = \hat{1} \, , \tag{10.11}$$

where $\hat{1}$ is the unit operator defined by

$$\hat{1}f = f \, . \tag{10.12}$$

Hermitian adjoint: (Hermitian conjugate)

Operator $\hat{A}^\dagger$ is the Hermitian adjoint of an operator $\hat{A}$ if for two functions f and g that vanish at infinity

$$\int f^* \hat{A} g dV = \int \left(\hat{A}^\dagger f \right)^* g dV \, . \tag{10.13}$$

There are several important properties of Hermitian conjugate.

Property 1: $\left(\hat{A}^\dagger \right)^\dagger = \hat{A}$.

Proof:

$$\int f^* \hat{A} g dV = \int \left(\hat{A}^\dagger f \right)^* g dV = \int g \left(\hat{A}^\dagger f \right)^* dV = \left(\int g^* \hat{A}^\dagger f dV \right)^*$$

$$= \left(\int \left((\hat{A}^\dagger)^\dagger g \right)^* f dV \right)^* = \int \left((\hat{A}^\dagger)^\dagger g \right) f^* dV$$

$$= \int f^* \left((\hat{A}^\dagger)^\dagger g \right) dV \, , \tag{10.14}$$

as required.

Property 2: $\left(\hat{A}\hat{B} \right)^\dagger = \hat{B}^\dagger \hat{A}^\dagger$.

Proof:

$$\int f^* \left(\hat{A}\hat{B} \right)^\dagger g dV = \int \left(\hat{A}\hat{B}f \right)^* g dV = \int \left[\hat{A} \left(\hat{B}f \right) \right]^* g dV. \tag{10.15}$$

Introducing the notation $\hat{B}f = u$, we get

$$\int \left(\hat{A}u \right)^* g dV = \int u^* \hat{A}^\dagger g dV = \int \left(\hat{B}f \right)^* \hat{A}^\dagger g dV$$

$$= \int f^* \hat{B}^\dagger \hat{A}^\dagger g dV \, . \tag{10.16}$$

Thus, $\left(\hat{A}\hat{B} \right)^\dagger = \hat{B}^\dagger \hat{A}^\dagger$, as required.

The above rules of operator algebra show that it is important to maintain the order of the factors because operator algebra is not in general commutative for multiplication. Many commonly used results in mathematics are generally not valid for operators since these are based on the assumption that multiplication is commutative. For example, the simple identity written for operators

$$e^{\hat{A}} e^{\hat{B}} = e^{(\hat{A}+\hat{B})} \tag{10.17}$$

is true only if $\hat{A}$ and $\hat{B}$ commute. The proof of this identity is left as a tutorial problem for the readers.

10.2 Hermitian Operators

We now turn to define a special class of operators called *Hermitian* operators. They are distinguished by the relationship of $\hat{A}$ and $\hat{A}^{\dagger}$. The Hermitian operators are of great importance in quantum physics, as they represent physical (measurable) quantities.

Definition

Operator $\hat{A}$ is called *Hermitian* if

$$\hat{A}^{\dagger} = \hat{A}, \tag{10.18}$$

i.e., when

$$\int f^{*} \hat{A} g dV = \int (\hat{A} f)^{*} g dV. \tag{10.19}$$

In other words, **an operator is Hermitian if the Hermitian conjugate is equal to the operator itself.**

10.2.1 *Properties of Hermitian operators*

There are several important properties of Hermitian operators.
    If $\hat{A}$ and $\hat{B}$ are Hermitian, then

Property 1. $\hat{A} + \hat{B}$ is Hermitian.

Property 2. $\hat{A}^{2}$, $\hat{A}^{3}$, etc. are Hermitian.

Property 3. $c\hat{A}$ is Hermitian if c is a real number.

Proof of property 3:

$$\int f^* \left(c\hat{A}\right) g dV = \int \left(c\hat{A}f\right)^* g dV = c^* \int \left(\hat{A}f\right)^* g dV \,,$$

$$(10.20)$$

as required.

Property 4. The product $\hat{A}\hat{B}$ of two Hermitian operators is Hermitian only if $\hat{A}$ and $\hat{B}$ commute.

Proof:

$$\left(\hat{A}\hat{B}\right)^\dagger = \hat{B}^\dagger \hat{A}^\dagger = \hat{B}\hat{A} \,. \qquad (10.21)$$

Hence, $(\hat{A}\hat{B})^\dagger \neq \hat{A}\hat{B}$, unless $\hat{A}$ and $\hat{B}$ commute, as required.

Property 5. From property 4, we find that the commutator $\left[\hat{A}, \hat{B}\right]$ is not Hermitian, even if $\hat{A}$ and $\hat{B}$ are Hermitian.

Proof:

$$\left[\hat{A}, \hat{B}\right]^\dagger = \left(\hat{A}\hat{B} - \hat{B}\hat{A}\right)^\dagger = \left(\hat{A}\hat{B}\right)^\dagger - \left(\hat{B}\hat{A}\right)^\dagger$$
$$= \hat{B}^\dagger \hat{A}^\dagger - \hat{A}^\dagger \hat{B}^\dagger = \hat{B}\hat{A} - \hat{A}\hat{B} = -\left[\hat{A}, \hat{B}\right] \,, \quad (10.22)$$

as required.

Example

Commutator of the position and momentum operators

$$[\hat{x}, \hat{p}_x] = i\hbar \,. \qquad (10.23)$$

The commutator of the two Hermitian operators is a complex number (nonphysical number).

Property 6. The commutator $i\left[\hat{A}, \hat{B}\right]$ is Hermitian.

The proof is left to the readers.

Property 7. For an arbitrary not necessary Hermitian operator $\hat{A}$, the product $\hat{A}\hat{A}^\dagger$ is Hermitian.

Proof: Take the Hermitian conjugate of $\hat{A}\hat{A}^\dagger$:

$$\left(\hat{A}\hat{A}^\dagger\right)^\dagger = \left(\hat{A}^\dagger\right)^\dagger \hat{A}^\dagger = \hat{A}\hat{A}^\dagger , \qquad (10.24)$$

as required.

Property 8. If $\hat{A}$ is non-Hermitian, $\hat{A} + \hat{A}^\dagger$ and $i\left(\hat{A} - \hat{A}^\dagger\right)$ are Hermitian. Hence, the operator $\hat{A}$ can be written as a linear combination of two Hermitian operators:

$$\hat{A} = \frac{1}{2}\left(\hat{A} + \hat{A}^\dagger\right) + \frac{1}{2i}\left(i\left(\hat{A} - \hat{A}^\dagger\right)\right) . \qquad (10.25)$$

10.2.2 *Examples of Hermitian Operators*

Example 1. Position operator $\hat{\vec{r}}$ is Hermitian.

Proof:
Since $|\vec{r}|$ is a real number and $\hat{\vec{r}}g$ is just a multiplication of the function g by the vector $\vec{r}$, we have

$$\int f^*\hat{\vec{r}}gdV = \int \hat{\vec{r}}f^*gdV = \int \left(\hat{\vec{r}}f\right)^* gdV . \qquad (10.26)$$

as required.

Example 2. Potential $\hat{V}(\vec{r})$ is Hermitian.

It is easy to prove. Since $\hat{\vec{r}}$ is Hermitian, an arbitrary function of $\vec{r}$ is also Hermitian.

Example 3. Momentum operator is Hermitian.

Proof:

$$\int_V \left(\hat{\vec{p}}f\right)^* gdV = \int_V \left(\frac{\hbar}{i}\nabla f\right)^* gdV = -\frac{\hbar}{i}\int_V (\nabla f^*)gdV$$

$$= -\frac{\hbar}{i}\int_V \nabla(f^*g)dV + \frac{\hbar}{i}\int_V f^*(\nabla g)dV$$

$$= -\frac{\hbar}{i}\int_V \nabla(f^*g)dV + \int_V f^*\left(\frac{\hbar}{i}\nabla g\right)dV. \quad (10.27)$$

Using Gauss's divergence theorem, we find that the above equation can be written as

$$= -\frac{\hbar}{i} \oint_S f^* g \, dS + \int_V f^* \left(\hat{p} g \right) dV \, . \tag{10.28}$$

First integral in Eq. (10.28) vanishes as f and g vanish at infinity, and therefore we get

$$\int_V \left(\hat{p} f \right)^* g \, dV = \int_V f^* \left(\hat{p} g \right) dV \, , \tag{10.29}$$

which means that $\hat{p}$ is Hermitian, as required.

10.3 Eigenvalues and Eigenvectors

We have defined before the eigenvalues and eigenfunctions of the Hamiltonian of a particle (see Section 7.2). The idea of eigenvalues and eigenfunctions can be extended to arbitrary operators. Thus, we can state: If

$$\hat{A} \Psi = \alpha \Psi \, , \tag{10.30}$$

then Ψ is an eigenfunction of $\hat{A}$ with eigenvalue α.

Example

Determine if the function $\Psi = e^{2x}$ is an eigenfunction of the operators (a) $\hat{A} = d/dx$, (b) $\hat{B} = ()^2$, and (c) $\hat{C} = \int dx$.

Solution

(a) Operating on the wave function Ψ with the operator $\hat{A}$, we obtain

$$\hat{A} \Psi = \frac{d}{dx} e^{2x} = 2e^{2x} = 2\Psi \, , \tag{10.31}$$

which is a constant times the original function. Therefore, $\Psi = e^{2x}$ is an eigenfunction of the operator $\hat{A} = d/dx$ with an eigenvalue $\alpha = 2$.

Solutions to the parts (b) and (c) are left to the readers.

10.3.1 *Eigenvalues of Hermitian Operator*

As we have already stated, Hermitian operators play an important role in quantum physics as they represent physical quantities (observables). We put this fact as a theorem.

Theorem: Eigenvalues of a Hermitian operator are real.

Proof:

Assume that α is an eigenvalue of a Hermitian operator $\hat{A}$ corresponding to the eigenfunction f, which is normalized and vanishes at infinity. Then

$$\alpha = \alpha \int_V |f|^2 dV = \int_V f^* \alpha f dV = \int_V f^* \hat{A} f dV$$

$$= \int_V \left(\hat{A} f\right)^* f dV = \int_V \alpha^* f^* f dV = \alpha^* \int_V |f|^2 dV$$

$$= \alpha^*, \tag{10.32}$$

as required.

10.4 Scalar Product and Orthogonality of Wave Functions

In this chapter, we will define a very important property of wave functions: **orthogonality**. This property, for example, allows to identify whether the given wave functions belong to the same operator or not.

We say that two functions $\Psi_1(\vec{r})$ and $\Psi_2(\vec{r})$ are orthogonal if

$$\int_{-\infty}^{+\infty} \Psi_1^*(\vec{r}) \Psi_2(\vec{r}) dV = 0. \tag{10.33}$$

The orthogonality condition of two functions is analogous to the orthogonality condition of two vectors. We know from basic mathematics that vectors are orthogonal when the scalar product of the vectors is zero. In analogy, we can write a scalar product of two functions as

$$(\Psi_i, \Psi_j) = \int \Psi_i^*(\vec{r}) \Psi_j(\vec{r}) dV = a_j \delta_{ij}, \tag{10.34}$$

where a_j is a positive constant and δ_{ij} is the Kronecker delta function.

When $a_j = 1$, we say that the functions are *orthonormal*.

The complex functions Ψ_i form a complex linear vector space. The infinite-dimensional vector space of orthonormal functions is called *Hilbert space*.

The scalar product

$$(\Psi_i, \Psi_i) = ||\Psi_i|| = \int |\Psi_i|^2 \, dV \,, \qquad (10.35)$$

where Ψ_i is a square integrable function, is called the *norm* of the state (vector) Ψ. For a state function that represents physical quantity, the norm is finite. If the functions are orthonormal, the norm $||\Psi_i|| = 1$.

Example of orthogonal functions: Typical examples of orthogonal functions are sine and cosine functions. Their product with any other function of the same class gives zero when integrated over all ranges of variable, unless the two multiplied functions are identical.

$$\int_0^{2\pi} \sin(m\phi)\sin(n\phi)\,d\phi = \begin{cases} 0 & \text{for } m \neq n \\ \pi & \text{for } m = n \end{cases} \qquad (10.36)$$

$$\int_0^{2\pi} \cos(m\phi)\cos(n\phi)\,d\phi = \begin{cases} 0 & \text{for } m \neq n \\ \pi & \text{for } m = n \end{cases} \qquad (10.37)$$

$$\int_0^{2\pi} \sin(m\phi)\cos(n\phi)\,d\phi = 0 \quad \text{for all } m \text{ and } n. \quad (10.38)$$

The above orthogonality properties are readily proved by direct integration.

From the orthogonality of the sine functions, we see that the eigenfunctions of a particle in an infinite square-well potential, Eq. (8.22), corresponding to different energies ($n \neq m$) are orthogonal.

Having available the definition of orthogonal functions, we can proceed to formulate an important property of eigenfunctions of a linear Hermitian operator.

Theorem: Eigenfunctions of a linear Hermitian operator belonging to different eigenvalues are orthogonal.

Proof:

Consider a Hermitian operator $\hat{A}$. Let f and g be two eigenfunctions of $\hat{A}$ corresponding to two different eigenvalues α_f and α_g, respectively:

$$\hat{A} f = \alpha_f f, \qquad \hat{A} g = \alpha_g g, \tag{10.39}$$

where α_f, α_g are real numbers.

Since $\left(\hat{A} f \right)^* = \alpha_f f^*$, we can write

$$\int \left(\hat{A} f \right)^* g \, dV - \int f^* \left(\hat{A} g \right) dV = (\alpha_f - \alpha_g) \int f^* g \, dV. \tag{10.40}$$

However

$$\int \left(\hat{A} f \right)^* g \, dV = \int f^* \left(\hat{A} g \right) dV, \tag{10.41}$$

and therefore the left-hand side of Eq. (10.40) vanishes. Since, $\alpha_f \neq \alpha_g$, we have

$$(f, g) = \int f^* g \, dV = 0, \tag{10.42}$$

as required.

10.5 Expectation Value of an Operator

In classical physics, an expectation or average or mean value of an arbitrary quantity A is obtained by weighting each measured value A_i by the associated probability P_i and summing over all the measurements N. Thus,

$$\langle A \rangle = \sum_i P_i A_i \qquad i = 1, 2, \ldots, N, \tag{10.43}$$

where P_i is a probability of measuring the value A_i.

How do we calculate expectation values in quantum physics?

Consider an operator $\hat{A}$ acting on a function Ψ_i. Suppose that $\hat{A} \Psi_i$ exists, then the scalar product

$$\left(\Psi_i, \hat{A} \Psi_i \right) = \int \Psi_i^* \hat{A} \Psi_i \, dV \tag{10.44}$$

is called the *expectation* or average or mean value of the operator $\hat{A}$ in the state Ψ_i.

Similarly, as in classical physics, the expectation value can be calculated from the probability density as

$$\langle \hat{A} \rangle = \int \hat{A}\rho(\vec{r})dV = \int \hat{A}\,|\Psi_i|^2\,dV = \int \Psi_i^* \hat{A}\Psi_i dV\,, \quad (10.45)$$

where the order of the factors under the integral is not important.

10.5.1 *Properties of Expectation Value*

Property 1. Expectation value of a Hermitian operator is real.

Proof:

$$\langle \hat{A} \rangle = \int \Psi_i^* \hat{A}\Psi_i dV = \int \left(\hat{A}\Psi_i \right)^* \Psi_i dV$$

$$= \int \Psi_i \left(\hat{A}\Psi_i \right)^* dV = \left(\int \Psi_i^* \hat{A}\Psi_i dV \right)^* = \langle \hat{A} \rangle^*\,, \quad (10.46)$$

as required.

Property 2. Expectation value of an arbitrary operator $\hat{B}$ satisfies the following equation of motion

$$\frac{d}{dt}\langle \hat{B} \rangle = \left\langle \frac{\partial \hat{B}}{\partial t} \right\rangle + \frac{i}{\hbar}\langle [\hat{H},\,\hat{B}] \rangle\,. \quad (10.47)$$

Proof:
Since

$$\langle \hat{B} \rangle = \int \Psi_i^* \hat{B}\Psi_i dV\,,$$

we have

$$\frac{d}{dt}\langle \hat{B} \rangle = \int \left(\frac{\partial \Psi^*}{\partial t} \right) \hat{B}\Psi dV + \int \Psi^* \left(\frac{\partial \hat{B}}{\partial t} \right) \Psi dV$$

$$+ \int \Psi^* \hat{B} \left(\frac{\partial \Psi}{\partial t} \right) dV\,.$$

From the Schrödinger equation,

$$i\hbar\frac{\partial \Psi}{\partial t} = \hat{H}\Psi\,,$$

and its complex conjugate

$$-i\hbar\frac{\partial\Psi^*}{\partial t} = \hat{H}\Psi^* \,,$$

we obtain

$$\frac{d}{dt}\langle\hat{B}\rangle = \left\langle\frac{\partial\hat{B}}{\partial t}\right\rangle + \frac{i}{\hbar}\int\hat{H}\Psi^*\hat{B}\Psi dV - \frac{i}{\hbar}\int\Psi^*\hat{B}\hat{H}\Psi dV \,.$$

Since $\hat{H}$ is Hermitian, we finally get

$$\frac{d}{dt}\langle\hat{B}\rangle = \left\langle\frac{\partial\hat{B}}{\partial t}\right\rangle + \frac{i}{\hbar}\int\Psi^*\left(\hat{H}\hat{B} - \hat{B}\hat{H}\right)\Psi dV$$

$$= \left\langle\frac{\partial\hat{B}}{\partial t}\right\rangle + \frac{i}{\hbar}\langle[\hat{H},\,\hat{B}]\rangle \,,$$

as required.

Thus, expectation value of the operator $\hat{B}$ can depend on time even if the operator does not depend explicitly on time $(\partial\hat{B}/\partial t = 0)$. When $[\hat{H},\,\hat{B}] = 0$, we have $d\langle\hat{B}\rangle/dt = 0$, and then the expectation value is constant in time. In analogy to classical physics, we call $\langle\hat{B}\rangle$ a constant of motion.

Worked Example

Calculate the expectation value of the x-coordinate of a particle in an energy state E_n of a one-dimensional well, discussed in Section 8.1.

Solution

The expectation value of x is

$$\langle x\rangle = \int\phi^*(x)x\phi(x)dx \,,$$

where $\phi(x)$ is the wave function of the particle.

Since

$$\phi(x) = \phi_n(x) = \sqrt{\frac{2}{a}}\sin\left(\frac{n\pi}{a}x\right) \,, \qquad \text{for} \quad 0 \le x \le a \,,$$

and $\phi(x)$ is zero for $x < 0$ and $x > a$, we obtain

$$\langle x \rangle = \int \phi^*(x) x \phi(x) dx = \frac{2}{a} \int_0^a dx \, x \sin^2 \left(\frac{n\pi}{a} x \right)$$

$$= \frac{1}{a} \int_0^a dx \, x \left[1 - \cos \left(\frac{2n\pi}{a} x \right) \right]$$

$$= \frac{1}{a} \int_0^a dx \left[x - x \cos \left(\frac{2n\pi}{a} x \right) \right]$$

$$= \frac{1}{a} \int_0^a dx \, x - \frac{1}{a} \int_0^a dx \, x \cos \left(\frac{2n\pi}{a} x \right)$$

$$= \frac{1}{a} \frac{1}{2} x^2 \Big|_0^a - \frac{1}{a} \left\{ -\frac{a^2}{(2n\pi)^2} \left[\cos \left(\frac{2n\pi}{a} x \right) \right. \right.$$
$$\left. \left. + \frac{2n\pi x}{a} \sin \left(\frac{2n\pi}{a} x \right) \right]_0^a \right\}.$$

Since $\cos(2n\pi) = \cos 0 = 1$, and $\sin(2n\pi) = \sin 0 = 0$, we find that

$$\langle x \rangle = \frac{1}{2} a \,,$$

independent of n! Physically, this value results from the fact that the wave function of the particle is symmetric about $x = a/2$ for all n. Note the expectation value is not equal to the most probable value, which is given by $|\phi(x)|^2$.

10.5.2 Useful General Property of Hermitian Operators

We have already shown that expectation values of Hermitian operators are real. In terms of the scalar product of two orthonormal functions, this is characterized by

$$\left(\Psi_i, \hat{A} \Psi_i \right) = \left(\Psi_i, \hat{A} \Psi_i \right)^* . \tag{10.48}$$

From this property, we have in general for Hermitian operators

$$\left(\Psi_i, \hat{A} \Psi_j \right) = \left(\Psi_j, \hat{A} \Psi_i \right)^* . \tag{10.49}$$

Proof:

$$\left(\Psi_i, \hat{A} \Psi_j \right) = \int \Psi_i^* \hat{A} \Psi_j dV = \int \left(\hat{A} \Psi_i \right)^* \Psi_j dV = \int \Psi_j \left(\hat{A} \Psi_i \right)^* dV$$

$$= \left(\int \Psi_j^* \hat{A} \Psi_i dV \right)^* = \left(\Psi_j, \hat{A} \Psi_i \right)^* , \quad \text{as required.}$$

The properties (10.48) and (10.49) are very often used to check whether operators are Hermitian. The following example will illustrate the procedure.

Worked Example

Consider two operators $\hat{A} = d/dx$ and $\hat{B} = d^2/dx^2$ acting on two orthonormal wave functions $\Psi_1 = a\sin(nx)$ and $\Psi_2 = a\cos(nx)$, where n is a real number, $a = 1/\sqrt{\pi}$ and $x \in \langle-\pi, \pi\rangle$.

Are the operators $\hat{A}$ and $\hat{B}$ Hermitian?

Solution

First, consider the operator $\hat{A} = d/dx$. Since

$$\hat{A}\Psi_1 = a\frac{d}{dx}\sin(nx) = an\cos(nx) = n\Psi_2 , \qquad (10.50)$$

and

$$\hat{A}\Psi_2 = a\frac{d}{dx}\cos(nx) = -an\sin(nx) = -n\Psi_1 , \qquad (10.51)$$

we find the following values of the scalar products

$$\begin{aligned}
\left(\Psi_1, \hat{A}\Psi_1\right) &= -n\left(\Psi_1, \Psi_2\right) = 0 , \\
\left(\Psi_1, \hat{A}\Psi_2\right) &= -n\left(\Psi_1, \Psi_1\right) = -n , \\
\left(\Psi_2, \hat{A}\Psi_2\right) &= -n\left(\Psi_2, \Psi_1\right) = 0 , \\
\left(\Psi_2, \hat{A}\Psi_1\right) &= n\left(\Psi_2, \Psi_2\right) = n .
\end{aligned} \qquad (10.52)$$

Hence

$$\left(\Psi_1, \hat{A}\Psi_1\right) = \left(\Psi_2, \hat{A}\Psi_2\right)^* , \qquad (10.53)$$

but

$$\left(\Psi_1, \hat{A}\Psi_2\right) = -an \neq \left(\Psi_2, \hat{A}\Psi_1\right)^* = an . \qquad (10.54)$$

Thus, the operator $\hat{A} = d/dx$ is not Hermitian.

Consider now the operator $\hat{B} = d^2/dx^2$. Since

$$\hat{B}\Psi_1 = a\frac{d^2}{dx^2}\sin(nx) = -an^2\sin(nx) = -n^2\Psi_1, \quad (10.55)$$

and

$$\hat{B}\Psi_2 = a\frac{d^2}{dx^2}\cos(nx) = -an^2\cos(nx) = -n^2\Psi_2, \quad (10.56)$$

we find the following values of scalar products

$$
\begin{aligned}
\left(\Psi_1, \hat{B}\Psi_1\right) &= -n^2\left(\Psi_1, \Psi_1\right) = -n^2, \\
\left(\Psi_1, \hat{B}\Psi_2\right) &= -n^2\left(\Psi_1, \Psi_2\right) = 0, \\
\left(\Psi_2, \hat{B}\Psi_2\right) &= -n^2\left(\Psi_2, \Psi_2\right) = -n^2, \\
\left(\Psi_2, \hat{B}\Psi_1\right) &= -n^2\left(\Psi_2, \Psi_1\right) = 0.
\end{aligned}
\quad (10.57)
$$

Hence

$$\left(\Psi_1, \hat{B}\Psi_1\right) = -n^2 = \left(\Psi_2, \hat{B}\Psi_2\right)^*, \quad (10.58)$$

and

$$\left(\Psi_1, \hat{B}\Psi_2\right) = 0 = \left(\Psi_2, \hat{B}\Psi_1\right)^*. \quad (10.59)$$

Thus, the operator $\hat{B} = d^2/dx^2$ is Hermitian.

> *I am not really here.*
>
> —Tim Allen

10.6 Heisenberg Uncertainty Principle Revisited

In Section 6.6, we have shown that the uncertainties in the position and momentum of a particle satisfy the relation

$$\Delta y \Delta p_y = h. \quad (10.60)$$

This relation says that the position and momentum of a particle cannot be measured simultaneously with the same precision. This is known as the **Heisenberg uncertainty relation**, or the Heisenberg uncertainty principle, and we will show that the relation is a direct consequence of the noncommutivity of the position and momentum operators

$$[\hat{y}, \hat{p}_y] = i\hbar. \tag{10.61}$$

In fact, the Heisenberg uncertainty relation can be formulated for arbitrary two Hermitian operators that do *not* commute. In other words, if $\hat{A}$ and $\hat{B}$ are two Hermitian operators that do not commute, the physical quantities represented by the operators cannot be measured simultaneously with the same precision.

Theorem: The variances $\langle(\Delta\hat{A})^2\rangle = \langle\hat{A}^2\rangle - \langle\hat{A}\rangle^2$ and $\langle(\Delta\hat{B})^2\rangle = \langle\hat{B}^2\rangle - \langle\hat{B}\rangle^2$ of two Hermitian operators satisfy the inequality

$$\langle(\Delta\hat{A})^2\rangle\langle(\Delta\hat{B})^2\rangle \geq -\frac{1}{4}\langle[\hat{A}, \hat{B}]\rangle^2, \tag{10.62}$$

which is called the *Heisenberg inequality*.

Proof: First, we prove that for an arbitrary operator $\hat{A}$, the following inequality holds:

$$\langle\hat{A}\hat{A}^\dagger\rangle \geq 0. \tag{10.63}$$

It is easy to prove the above inequality using the definition of the expectation value:

$$\langle\hat{A}\hat{A}^\dagger\rangle = \int \Psi^*\hat{A}\hat{A}^\dagger\Psi dV = \int \left(\hat{A}^\dagger\Psi\right)^*\hat{A}^\dagger\Psi dV = \int \left|\hat{A}^\dagger\Psi\right|^2 dV \geq 0. \tag{10.64}$$

Now we prove that for two Hermitian operators, the following inequality is satisfied:

$$\langle\hat{A}^2\rangle\langle\hat{B}^2\rangle \geq -\frac{1}{4}\langle[\hat{A}, \hat{B}]\rangle^2. \tag{10.65}$$

To prove it, we introduce an operator

$$\hat{D} = \hat{A} + iz\hat{B}, \tag{10.66}$$

where z is an arbitrary real number. Hence, from Eq. (10.63), we find

$$\langle\hat{D}\hat{D}^\dagger\rangle = \langle(\hat{A} + iz\hat{B})(\hat{A} - iz\hat{B})\rangle$$
$$= \langle\hat{A}^2\rangle - iz(\langle\hat{A}\hat{B} - \hat{B}\hat{A}\rangle) + z^2\langle\hat{B}^2\rangle \geq 0. \tag{10.67}$$

This inequality is satisfied when

$$-\langle\hat{A}\hat{B} - \hat{B}\hat{A}\rangle^2 - 4\langle\hat{A}^2\rangle\langle\hat{B}^2\rangle \leq 0. \tag{10.68}$$

Hence

$$\langle\hat{A}^2\rangle\langle\hat{B}^2\rangle \geq -\frac{1}{4}\langle[\hat{A}, \hat{B}]\rangle^2, \tag{10.69}$$

as required.

Finally, since

$$[\Delta\hat{A}, \Delta\hat{B}] = [\hat{A}, \hat{B}], \tag{10.70}$$

where $\Delta\hat{U} = \hat{U} - \langle\hat{U}\rangle$, $(\hat{U} = \hat{A}, \hat{B})$, and replacing in Eq. (10.69), $\hat{A} \rightarrow \Delta\hat{A}$ and $\hat{B} \rightarrow \Delta\hat{B}$, we obtain the Heisenberg uncertainty relation (10.62), as required.

Some of the practical applications of the Heisenberg uncertainty relation are illustrated in the following examples.

Worked Example

The Heisenberg uncertainty relation for the position and momentum operators.

Since the position and momentum operators satisfy the commutation relation

$$[\hat{x}, \hat{p}_x] = i\hbar, \tag{10.71}$$

we obtain by substituting into Eq. (10.62), $\hat{A} = \hat{x}$ and $\hat{B} = \hat{p}_x$

$$\langle(\Delta\hat{x})^2\rangle\langle(\Delta\hat{p}_x)^2\rangle \geq \frac{1}{4}\hbar^2, \tag{10.72}$$

or in terms of the standard deviations (fluctuations)

$$\delta x \delta p_x \geq \frac{1}{2}\hbar, \tag{10.73}$$

where $\delta x = \sqrt{\langle(\Delta\hat{x})^2\rangle}$ and $\delta p_x = \sqrt{\langle(\Delta\hat{p}_x)^2\rangle}$.

Similarly, we can show that for the y- and z-components of the position and momentum

$$\delta y \delta p_y \geq \frac{1}{2}\hbar \quad \text{and} \quad \delta z \delta p_z \geq \frac{1}{2}\hbar. \tag{10.74}$$

Note that Eq. (10.60) satisfies the Heisenberg inequality as $h > \hbar/2$.

The uncertainty relations (10.73) and (10.74) show one of the strange quantum behavior. The particle can never truly be at rest. Even in its lowest energy state, at a temperature of absolute zero, its position and momentum are still subject to fluctuations, called quantum fluctuations.

Worked Example

The Heisenberg uncertainty relation for the components of the electron spin.

Since the components of the electron spin satisfy the commutation relation[a]

$$[\hat{\sigma}_x, \hat{\sigma}_y] = 2i\hat{\sigma}_z, \tag{10.75}$$

where $\hat{\sigma}_x, \hat{\sigma}_y, \hat{\sigma}_z$ are the operators corresponding to the three components of the electron spin, we obtain

$$\langle(\Delta\hat{\sigma}_x)^2\rangle\langle(\Delta\hat{\sigma}_y)^2\rangle \geq \langle\hat{\sigma}_z\rangle^2, \tag{10.76}$$

or

$$\delta\sigma_x\delta\sigma_y \geq |\langle\hat{\sigma}_z\rangle|. \tag{10.77}$$

The uncertainty relation (10.77) shows that the components of the electron spin cannot be measured simultaneously with the same precision. Again, we face a strange quantum behavior. The spin of the electron, even at a temperature of absolute zero, exhibits quantum fluctuations.

10.7 Expansion of Wave Functions in the Basis of Orthonormal Functions

Orthonormal wave functions are very useful in quantum physics, in particular those of Hermitian operators. They can be employed to represent an arbitrary wave function. In other words, we may expand an arbitrary wave function in the basis of orthonormal functions. This is a very useful property of orthonormal functions, which, similar to the orthogonality, arises from the properties of vectors. The reason of using expansions is that it is more convenient to perform any calculations and mathematical operations on orthonormal functions, whose properties are completely known, rather than on non-orthogonal or not completely determined functions.

---

[a]The proof of the commutation relation is left to the readers as a tutorial problem.

To illustrate this procedure, consider a simple example: In the Cartesian coordinates, an arbitrary vector $\vec{A}$ can be written as a linear combination of the orthonormal unit vectors

$$\vec{A} = (\vec{i} \cdot \vec{A})\vec{i} + (\vec{j} \cdot \vec{A})\vec{j} + (\vec{k} \cdot \vec{A})\vec{k}, \tag{10.78}$$

where $\vec{i}$, $\vec{j}$, and $\vec{k}$ are unit vectors in the directions x, y, and z, respectively.

Proof: We know from the vector analysis that in the Cartesian coordinates, an arbitrary vector $\vec{A}$ may be presented in terms of components A_x, A_y, A_z, and three unit vectors oriented in the directions of the coordinate axis

$$\vec{A} = A_x\vec{i} + A_y\vec{j} + A_z\vec{k}. \tag{10.79}$$

Since the components are the projections of $\vec{A}$ on the coordinate axis

$$A_x = \vec{i} \cdot \vec{A}, \qquad A_y = \vec{j} \cdot \vec{A}, \qquad A_z = \vec{k} \cdot \vec{A}, \tag{10.80}$$

we find that the vector (10.79) can be written in the form

$$\vec{A} = (\vec{i} \cdot \vec{A})\vec{i} + (\vec{j} \cdot \vec{A})\vec{j} + (\vec{k} \cdot \vec{A})\vec{k}, \tag{10.81}$$

as required.

We can extend this property to m-dimensional space and state that **an arbitrary vector $\vec{A}$ can be written as a linear combination of the coordinate (basis) orthogonal unit vectors $\vec{e}$ as**

$$\vec{A} = \left(\vec{e}_1 \cdot \vec{A}\right)\vec{e}_1 + \left(\vec{e}_2 \cdot \vec{A}\right)\vec{e}_2 + \ldots + \left(\vec{e}_m \cdot \vec{A}\right)\vec{e}_m = \sum_{n=1}^{m} \left(\vec{e}_n \cdot \vec{A}\right)\vec{e}_n, \tag{10.82}$$

where $(\vec{e}_n \cdot \vec{A})$ is the scalar product of $\vec{e}_n$ and $\vec{A}$, (nth component of $\vec{A}$), and $\vec{e}_i \cdot \vec{e}_j = \delta_{ij}$.

The norm (magnitude) of the vector $\vec{A}$ is

$$||\vec{A}||^2 = |\vec{A}| = \sum_n \left(\vec{e}_n \cdot \vec{A}\right)^2. \tag{10.83}$$

Thus, we see that an arbitrary vector can be expressed as a linear combination of the orthonormal vectors $\vec{e}_n$.

The same ideas carry over to the space of wave functions. Namely, **an arbitrary wave function Ψ can be expanded in terms of orthonormal wave functions Ψ_n as**

$$\Psi = \sum_n c_n \Psi_n, \tag{10.84}$$

(discrete spectrum of Ψ_n), or in the case of a continuous spectrum of Ψ_n

$$\Psi(\vec{r}) = \int c_n(\vec{r})\Psi_n(\vec{r})dV_n , \qquad (10.85)$$

where c_n are arbitrary (unknown) expansion coefficients, and dV_n is the volume element of the space the orthonormal functions $\Psi_n(\vec{r})$ are spanned.

We can find the coefficients $c_n(\vec{r})$ by multiplying Eq. (10.85) by $\Psi_m^*(\vec{r})$ and integrating over all space as follows:

$$\int \Psi_m^*(\vec{r})\Psi(\vec{r})dV = \int c_n(\vec{r})\int \Psi_m^*(\vec{r})\Psi_n(\vec{r})dVdV_n = c_m(\vec{r}) , \qquad (10.86)$$

where we have used the orthonormality property of the Ψ_n functions

$$\int \Psi_m^*(\vec{r})\Psi_n(\vec{r})dV_n = \delta_{nm} . \qquad (10.87)$$

In general, the coefficients $c_m(\vec{r})$ are complex numbers and are called the components of the function Ψ in the basis of the orthonormal functions Ψ_m. The components determine the function completely, and very often the coefficients $c_m(\vec{r})$ are called a representation of the wave function Ψ in the basis Ψ_m.

The coefficients $c_m(\vec{r})$ satisfy the following relation:

$$\int_V |c_m(\vec{r})|^2 dV = 1 . \qquad (10.88)$$

Proof: Multiplying Eq. (10.85) by $\Psi^*(\vec{r})$ and integrating over V, we obtain

$$\int_V \Psi^*(\vec{r})\Psi(\vec{r})dV = \int_V |\Psi(\vec{r})|^2 dV$$

$$= \int_V \int_{V_n} \int_{V_m} c_m^*(\vec{r})c_n(\vec{r})\Psi_n\Psi_m dV_n dV_m dV$$

$$= \int_V \int_{V_n} c_m^*(\vec{r})c_n(\vec{r})\delta_{mn}dV_n dV = \int_V |c_m(\vec{r})|^2 dV . \quad (10.89)$$

Since, $\int_V |\Psi|^2 dV = 1$, we get $\int_V |c_m(\vec{r})|^2 dV = 1$, as required.

From Eqs. (10.85) and (10.88), we see that $|c_m(\vec{r})|^2$ can be interpreted as the probability that a system, described by the wave function $\Psi(\vec{r})$, is in the state described by the wave function $\Psi_m(\vec{r})$.

Example

Let $\hat{A}$ be an operator and $\Psi(\vec{r})$ be an unknown wave function that is not an eigenfunction of $\hat{A}$. Suppose that we cannot find the explicit form of $\Psi(\vec{r})$. However, we can find a form of the wave function in the basis of the eigenfunctions of $\hat{A}$. If $\Psi_m(\vec{r})$ is an eigenfunction of $\hat{A}$, then

$$\Psi(\vec{r}) = \int c_m(\vec{r})\Psi_m(\vec{r})dV_m . \qquad (10.90)$$

Tutorial Problems

Problem 10.1 Show that three arbitrary operators $\hat{A}$, $\hat{B}$, and $\hat{C}$ satisfy the following identities:

$$[\hat{A} + \hat{B}, \hat{C}] = [\hat{A}, \hat{C}] + [\hat{B}, \hat{C}],$$
$$[\hat{A}, \hat{B}\hat{C}] = [\hat{A}, \hat{B}]\hat{C} + \hat{B}[\hat{A}, \hat{C}].$$

Problem 10.2 Let $\hat{A}$, $\hat{B}$, $\hat{C}$ are arbitrary linear operators. Prove that

(a) $\left[\hat{A}\hat{B}, \hat{C}\right] = \left[\hat{A}, \hat{C}\right]\hat{B} + \hat{A}\left[\hat{B}, \hat{C}\right]$,
(b) $\left[\hat{A}, \left[\hat{B}, \hat{C}\right]\right] + \left[\hat{B}, \left[\hat{C}, \hat{A}\right]\right] + \left[\hat{C}, \left[\hat{A}, \hat{B}\right]\right] = 0$.

Problem 10.3 Let

$$\left[\hat{A}, \hat{B}\right] = i\hbar\hat{C} \quad \text{and} \quad \left[\hat{B}, \hat{C}\right] = i\hbar\hat{A} .$$

Show that

$$\hat{B}\left(\hat{C} + i\hat{A}\right) = \left(\hat{C} + i\hat{A}\right)\left(\hat{B} + \hbar\right),$$
$$\hat{B}\left(\hat{C} - i\hat{A}\right) = \left(\hat{C} - i\hat{A}\right)\left(\hat{B} - \hbar\right) .$$

Problem 10.4 *Taylor expansion*
Show that

$$e^{\hat{A}}\hat{B}e^{-\hat{A}} = \hat{B} + \frac{1}{1!}\left[\hat{A}, \hat{B}\right] + \frac{1}{2!}\left[\hat{A}, \left[\hat{A}, \hat{B}\right]\right]$$
$$+ \frac{1}{3!}\left[\hat{A}, \left[\hat{A}, \left[\hat{A}, \hat{B}\right]\right]\right] + \dots$$

This formula shows that the calculation of complicated exponential-type operator functions can be simplified to the calculation of a series of commutators.

Problem 10.5 Consider two arbitrary operators $\hat{A}$ and $\hat{B}$. If $\hat{A}$ commutes with their commutator, $[\hat{A}, \hat{B}]$:

(a) Prove that for a positive integer n,

$$[\hat{A}^n, \hat{B}] = n\hat{A}^{n-1}[\hat{A}, \hat{B}] .$$

(b) Apply the commutation relation (a) to the special case of $\hat{A} = \hat{x}$, $\hat{B} = \hat{p}_x$, and show that

$$[f(\hat{x}), \hat{p}_x] = i\hbar\frac{df}{dx} ,$$

assuming that $f(\hat{x})$ can be expanded in a power series of the operator $\hat{x}$.

Problem 10.6 Prove that the *Campbell–Baker–Hausdorff* operator identity

$$e^{\hat{A}+\hat{B}} = e^{\hat{A}}e^{\hat{B}}e^{-\frac{1}{2}[\hat{A},\hat{B}]}$$

is valid only for two operators satisfying the commutation relations

$$[\hat{A}, [\hat{A}, \hat{B}]] = [\hat{B}, [\hat{A}, \hat{B}]] = 0 ,$$

i.e., when each of the operators commutes with their commutator.

Problem 10.7 Determine if the function $\phi = e^{ax}\sin x$, where a is a real constant, is an eigenfunction of the operator d/dx and d^2/dx^2. If it is, determine any eigenvalue.

Problem 10.8 What are the eigenvalues and eigenfunctions of the operator $(id/dx)^2$ if the eigenfunctions are required to be zero when $x = 0$ and 2?

Problem 10.9 Calculate the expectation value of the x-coordinate of a particle in an energy state E_n of a one-dimensional box.

Problem 10.10 Prove, using the condition (10.49) and the wave functions Ψ_1 and Ψ_2 of the above example, that the momentum operator $\hat{p}_x = -i\hbar d/dx$ is Hermitian.

Problem 10.11 For a particle in an infinite square-well potential represented by the position $\hat{x}$ and momentum $\hat{p}_x$ operators, check the uncertainty principle $\Delta x \Delta p_x \geq \hbar/2$ for $n = 1$, where $\Delta x = \sqrt{\langle \hat{x}^2 \rangle - \langle \hat{x} \rangle^2}$ and $\Delta p_x = \sqrt{\langle \hat{p}_x^2 \rangle - \langle \hat{p}_x \rangle^2}$.

Problem 10.12 The expectation value of an arbitrary operator $\hat{A}$ in the state $\phi(x)$ is given by

$$\langle \hat{A} \rangle = \int \phi^*(x) \hat{A} \phi(x) dx .$$

(a) Calculate expectation values (i) $\langle \hat{x} \hat{p}_x \rangle$, (ii) $\langle \hat{p}_x \hat{x} \rangle$, and (iii) $(\langle \hat{x} \hat{p}_x \rangle + \langle \hat{p}_x \hat{x} \rangle)/2$ of the product of position ($\hat{x} = x$) and momentum ($\hat{p}_x = -i\hbar \frac{d}{dx}$) operators of a particle represented by the wave function

$$\phi(x) = \sqrt{\frac{2}{a}} \sin\left(\frac{\pi x}{a}\right) ,$$

where $0 \leq x \leq a$.

(b) The operators $\hat{x}$ and $\hat{p}_x$ are Hermitian. Which of the products (i) $\hat{x} \hat{p}_x$, (ii) $\hat{p}_x \hat{x}$, and (iii) $(\hat{x} \hat{p}_x + \hat{p}_x \hat{x})/2$ are Hermitian?

(c) Explain which of the results of (a) are acceptable as the expectation values of physical quantities.

Chapter 11

Dirac Bra-Ket Notation

Dirac introduced a very useful (compact) notation of state vectors (wave functions) Ψ_i in terms of **"bra"** $\langle i|$ and **"ket"** $|i\rangle$ vectors. This notation allows to make the formal expressions of quantum physics more transparent and easier to manipulate.

For example, a wave function Ψ_i can be expressed by a ket vector $|\Psi_i\rangle$, and Ψ_i^* by a bra vector $\langle \Psi_i|$. This notation can be further simplified to $|i\rangle$ and $\langle i|$, respectively.

Let us illustrate what kind of simplifications we will get using the Dirac notation.

In the Dirac notation, a scalar product is written as

$$\left(\Psi_i, \Psi_j\right) = \langle \Psi_i|\Psi_j\rangle = \langle i|j\rangle, \tag{11.1}$$

which is called a *bracket*.

We note in the bracket expression the double vertical bar is dropped in favor of a single one.

For orthonormal vectors, we have used the notation $\left(\Psi_i, \Psi_j\right) = \delta_{ij}$, which in the Dirac notation takes the form $\langle i|j\rangle = \delta_{ij}$.

Since $\left(\Psi_i, \Psi_j\right) = \left(\Psi_j, \Psi_i\right)^*$, we have in the Dirac notation $\langle i|j\rangle = \langle j|i\rangle^*$.

In the bra-ket notation, the definition of the Hermitian adjoint becomes

$$\langle i|\hat{A}|j\rangle = \left(\langle j|\hat{A}^\dagger|i\rangle\right)^*, \tag{11.2}$$

Quantum Physics for Beginners
Zbigniew Ficek
Copyright © 2016 Pan Stanford Publishing Pte. Ltd.
ISBN 978-981-4669-38-2 (Hardcover), 978-981-4669-39-9 (eBook)
www.panstanford.com

or

$$\langle j|\hat{A}^\dagger|i\rangle = \langle i|\hat{A}|j\rangle^* \, . \tag{11.3}$$

Thus, for a Hermitian operator,

$$\langle i|\hat{A}|j\rangle = \langle j|\hat{A}|i\rangle^* \, . \tag{11.4}$$

Expectation value of an operator $\hat{A}$ in a state $|i\rangle$ is given by $\langle i|\hat{A}|i\rangle$. We write a linear superposition of ket states as

$$|a\rangle = \sum_n \lambda_n |n\rangle \, , \qquad \text{discrete states} \tag{11.5}$$

or

$$|a\rangle = \int \lambda(x)|x\rangle dx \, . \qquad \text{continuous states} \tag{11.6}$$

The bra-ket notation also extends to action of operators on state vectors.

A linear operator $\hat{A}$ associates with every ket $|i\rangle$ another ket $|j\rangle$:

$$\hat{A}|i\rangle = |j\rangle \, , \tag{11.7}$$

such that

$$\hat{A}\left(|a\rangle + |b\rangle\right) = \hat{A}|a\rangle + \hat{A}|b\rangle \, , \quad \text{and} \quad \hat{A}\lambda|a\rangle = \lambda\left(\hat{A}|a\rangle\right) , \tag{11.8}$$

where λ is a number.

Hermitian conjugate of $\hat{A}|i\rangle$ is $\langle i|\hat{A}^\dagger$.

An arbitrary ket state $|a\rangle$ can be expanded in terms of orthonormal ket states as

$$|a\rangle = \sum_n c_n |n\rangle \, . \tag{11.9}$$

Since $|n\rangle$ are orthonormal ($\langle m|n\rangle = \delta_{mn}$), we get for c_n:

$$\langle n|a\rangle = \sum_m c_m \langle n|m\rangle = \sum_m c_m \delta_{nm} = c_n \, . \tag{11.10}$$

Thus, we can write the ket state $|a\rangle$ as

$$|a\rangle = \sum_n |n\rangle\langle n|a\rangle \, , \tag{11.11}$$

from which we find a useful property of the orthonormal ket states

$$\sum_n |n\rangle\langle n| = \hat{1} \, , \tag{11.12}$$

where $\hat{1}$ is the unit operator.

The product ket-bra ($|n\rangle\langle n|$) is called a *projection operator,* and the relation (11.12) is called the *completeness relation.*

The product $|n\rangle\langle n|$ is not a product of two functions in the normal sense. It is a *dyadic* product: the function $|n\rangle$ "stands" at the function $\langle n|$.

The completeness relation is very useful in quantum physics as it provides significant simplifications in calculations involving operators and state vectors. Examples will be presented in Section 11.2; see also Tutorial Problem 11.1.

11.1 Projection Operator

In general, we can define projection operator of two different bra-ket states as

$$\hat{P}_{mn} = |m\rangle\langle n| . \tag{11.13}$$

The term "projection" results from a specific property of this operator, that it projects an arbitrary state vector $|a\rangle$ onto the ket state $|m\rangle$:

$$\hat{P}_{mn}|a\rangle = |m\rangle\langle n|a\rangle , \tag{11.14}$$

i.e., the result of its action is the state $|m\rangle$ with the amplitude $\langle n|a\rangle$.

When $m = n$, the projection operator reduces to a diagonal form $\hat{P}_{nn}$, which satisfies the relation

$$\hat{P}_{nn}^2 = \hat{P}_{nn} . \tag{11.15}$$

It is easy to prove. Since $\langle n|n\rangle = 1$, we have

$$\hat{P}_{nn}^2 = \hat{P}_{nn}\hat{P}_{nn} = |n\rangle\langle n|n\rangle\langle n| = \hat{P}_{nn} . \tag{11.16}$$

Thus, **the square of $\hat{P}_{nn}$ equals itself**.

Another important property of the projection operator:

The operator $\hat{P}_{nn}$ **is Hermitian, but** $\hat{P}_{mn}$, $(m \neq n)$ **is not Hermitian**.

The proof is straightforward. First note that

$$\langle i|\hat{P}_{mn}|j\rangle = \langle i|m\rangle\langle n|j\rangle = \delta_{im}\delta_{nj} . \tag{11.17}$$

Then, we have $\langle m|\hat{P}_{mn}|n\rangle = 1$. However, $\langle n|\hat{P}_{mn}|m\rangle = 0$, and then

$$\langle m|\hat{P}_{mn}|n\rangle \neq \left(\langle n|\hat{P}_{mn}|m\rangle\right)^* . \tag{11.18}$$

Thus, the operator $\hat{P}_{mn}$, $(m \neq n)$ is not Hermitian.

11.2 Representations of Linear Operators

The projection operator and the completeness relation are very important quantities in quantum physics. They are very useful in calculations and are often employed to represent an arbitrary operator and a wave function (state) in terms of orthonormal states. As we have already seen, the reason of employing representations is that it is more convenient to perform any calculations and mathematical operations on orthonormal functions, whose properties are completely known, rather than on non-orthogonal or not completely determined functions. To illustrate the procedure of representations, we will consider an arbitrary operator $\hat{A}$ and will show how to represent the operator in terms of projection operators of known orthonormal states $|m\rangle$.

To arrive at a representation of the operator $\hat{A}$ in the basis of the orthonormal states $|m\rangle$, we make use of the completeness relation by multiplying the operator $\hat{A}$ on both sides by unity in the form

$$1 = \sum_m |m\rangle\langle m| . \qquad (11.19)$$

This gives

$$\hat{A} = \left(\sum_m |m\rangle\langle m| \right) \hat{A} \left(\sum_n |n\rangle\langle n| \right) = \sum_{m,n} |m\rangle\langle m|\hat{A}|n\rangle\langle n|$$

$$= \sum_{m,n} \langle m|\hat{A}|n\rangle |m\rangle\langle n| = \sum_{m,n} A_{mn} \hat{P}_{mn} , \qquad (11.20)$$

where $A_{mn} = \langle m|\hat{A}|n\rangle$ are numbers, which are interpreted as matrix elements of the operator $\hat{A}$ in the basis of the states $|m\rangle$.

Thus, this simple application of the completeness relation shows that an arbitrary operator can be written (represented) as a linear combination of projection operators $\hat{P}_{mn}$ of the known orthonormal states $|m\rangle$. The following example illustrates the concept just introduced.

Worked Example

Consider an operator $\hat{A} = d^2/dx^2$. Write the operator in terms of projection operators involving the two lowest energy states of a particle in an infinite well potential.

Solution

We have found before in Section 8.1.2 that the two lowest energy states of a particle in an infinite well potential are orthonormal and are given by

$$|1\rangle = \phi_1(x) = \sqrt{\frac{2}{a}} \sin\left(\frac{\pi}{a}x\right),$$

$$|2\rangle = \phi_2(x) = \sqrt{\frac{2}{a}} \sin\left(\frac{2\pi}{a}x\right).$$

According to Eq. (11.20), an arbitrary operator $\hat{A}$ can be represented in the basis $|1\rangle$, $|2\rangle$ as

$$\hat{A} = A_{11}|1\rangle\langle 1| + A_{12}|1\rangle\langle 2| + A_{21}|2\rangle\langle 1| + A_{22}|2\rangle\langle 2|,$$

where $A_{mn} = \langle m|\hat{A}|n\rangle$, $(m, n = 1, 2)$ are the matrix elements of the operator. Since

$$\hat{A}|1\rangle = \frac{d^2}{dx^2}\sqrt{\frac{2}{a}} \sin\left(\frac{\pi}{a}x\right) = -\left(\frac{\pi}{a}\right)^2 |1\rangle,$$

$$\hat{A}|2\rangle = \frac{d^2}{dx^2}\sqrt{\frac{2}{a}} \sin\left(\frac{2\pi}{a}x\right) = -\left(\frac{2\pi}{a}\right)^2 |2\rangle,$$

and the states $|1\rangle$ and $|2\rangle$ are orthonormal, we easily find that

$$A_{11} = -\left(\frac{\pi}{a}\right)^2, \qquad A_{22} = -\left(\frac{2\pi}{a}\right)^2, \qquad A_{12} = A_{21} = 0.$$

Thus, in the basis of the orthonormal states $|1\rangle$, $|2\rangle$, the operator $\hat{A}$ is represented in terms of the projection operators as

$$\hat{A} = -\left(\frac{\pi}{a}\right)^2 (|1\rangle\langle 1| + 4|2\rangle\langle 2|).$$

11.3 Representations of an Expectation Value

We have learned that an arbitrary state $|a\rangle$ can be expanded in terms of orthonormal states $|n\rangle$:

$$|a\rangle = \sum_n c_n|n\rangle. \tag{11.21}$$

We will apply this expansion to find a representation of an expectation value $\langle \hat{A} \rangle$ of an operator $\hat{A}$ in terms of the orthonormal states $|m\rangle$. Using Eq. (11.21), we can write the expectation value as

$$\langle \hat{A} \rangle = \langle a|\hat{A}|a \rangle = \sum_{m,n} \langle n|\hat{A}|m \rangle c_n^* c_m . \tag{11.22}$$

If $|m\rangle$ is an eigenfunction of $\hat{A}$, i.e., $\hat{A}|m\rangle = A_m|m\rangle$, then

$$\langle \hat{A} \rangle = \sum_{m,n} c_n^* c_m A_m \delta_{mn} = \sum_n A_n |c_n|^2 . \tag{11.23}$$

Thus, the modulus square of the expansion coefficients is the probability that the quantity described by the operator $\hat{A}$ is in the state $|n\rangle$.

As $\langle \hat{A} \rangle$ is a weighted sum of the eigenvalues, this suggests that the eigenvalues represent the possible results of measurement, while $|c_n|^2$ is the probability that the eigenvalue A_n will be obtained as the result of any individual measurement.

Thus, in quantum physics, even if a given system is in its eigenstate, the measurement is not certain.

This is in contrast to classical physics. In classical physics, the measurement of a physical quantity at any time always leads to a definite result. In quantum physics, the measurement of the physical quantity at any time leads to a range of possible results, each occurring with a certain probability. In this sense, quantum physics is probabilistic.

Results of any measurement in physics are real numbers. Since eigenvalues of Hermitian operators are real, we *postulate* that every physical quantity that is measurable is specified in quantum physics by a linear Hermitian operator $\hat{A}$, which is also called an *observable*.

In quantum physics, the set of possible measured values for a physical quantity is the set of eigenvalues of a linear Hermitian operator specifying the physical quantity.

Example

As an example of this procedure, let us consider a particle specified by a wave function Ψ_a, or in the Dirac notation, by $|a\rangle$. Let $\hat{H}$

be the Hamiltonian (energy) of the particles and $|n\rangle$ are known eigenfunctions of $\hat{H}$.

If $|a\rangle$ is an eigenfunction of $\hat{H}$, then

$$\hat{H}|a\rangle = E_a|a\rangle, \tag{11.24}$$

where E_a is the eigenvalue (energy) of the particle. Thus, $E_a = \langle a|\hat{H}|a\rangle$.

If $|a\rangle$ is not an eigenfunction of $\hat{H}$, then we can expand $|a\rangle$ in terms of the eigenfunctions $|n\rangle$ as

$$|a\rangle = \sum_n c_n|n\rangle, \tag{11.25}$$

and find that

$$\langle a|\hat{H}|a\rangle = \sum_n E_n |c_n|^2. \tag{11.26}$$

Hence, the measurement of energy of the particle in the state $|a\rangle$ leads to a range of possible results, each occurring with probability $|c_n|^2$. Thus, $|c_n|^2$ is the probability that the measurement of $\hat{H}$ will give the value E_n.

Since, $|a\rangle = \sum c_n|n\rangle$, we say that the state of the particle is a superposition of the eigenfunctions of $\hat{H}$.

Tutorial Problems

Problem 11.1 *Useful application of the completeness relation*
As we have mentioned in the chapter, the completeness relation is very useful in calculations involving operators and state vectors. Consider the following example.

Let A_{il} and B_{lj} be matrix elements of two arbitrary operators $\hat{A}$ and $\hat{B}$ in a basis of orthonormal vectors. Show, using the completeness relation, that matrix elements of the product operator $\hat{A}\hat{B}$ in the same orthonormal basis can be found from the multiplication of the matrix elements A_{il} and B_{lj} as

$$\left(\hat{A}\hat{B}\right)_{ij} = \sum_{l=1}^{n} A_{il} B_{lj}.$$

Problem 11.2 *Eigenvalues of the projection operator*

Show that the eigenvalues of the projection operator P_{nn} are 0 or 1.

Problem 11.3 *Sum of two projection operators*

Let P_{nn} and P_{mm} be projection operators. Show that the sum $P_{nn} + P_{mm}$ is a projection operator if and only if $P_{nn} P_{mm} = 0$.

Chapter 12

Matrix Representations

We have already learned that an arbitrary operator can be represented in a basis of orthonormal states as a linear combination of projection operators. In this chapter, we will illustrate how one can obtain a matrix representation of a state vector, an operator, and an eigenvalue equation in a basis of orthonormal states. We will learn that matrix representations are very useful in calculations of the properties of operators, because algebra of matrices is very simple and completely developed. Readers familiar with the algebra of matrices agree that mathematical operations on matrices such as multiplication by a scalar, addition, subtraction, multiplication and diagonalization are simple and easy to perform. In fact, the application of the algebra of matrices to quantum physics results from the direct relations between operators and matrices that any mathematical relation that holds between operators also holds between the corresponding matrices. The eigenvalues and eigenvectors of the matrix representing a given operator $\hat{A}$ in the basis of orthonormal states are the same as the eigenvalues and eigenvectors of $\hat{A}$.

Quantum Physics for Beginners
Zbigniew Ficek
Copyright © 2016 Pan Stanford Publishing Pte. Ltd.
ISBN 978-981-4669-38-2 (Hardcover), 978-981-4669-39-9 (eBook)
www.panstanford.com

12.1 Matrix Representation of State Vectors

Using an orthonormal basis, we can represent an arbitrary normalized state vector $|a\rangle$ as a linear superposition of the basis states

$$|a\rangle = \sum_n c_n |n\rangle , \tag{12.1}$$

where, in general, the coefficients c_n are complex numbers, and $\sum_n |c_n|^2 = 1$.

The set of the expansion coefficients $c_1, c_2, \ldots$ defines the state $|a\rangle$ and is called the *representation* of $|a\rangle$ in the basis of the orthonormal states $|n\rangle$.

We can write the set of the coefficients c_n as a column (ket) vector

$$|a\rangle = \begin{pmatrix} c_1 \\ c_2 \\ \cdot \\ \cdot \\ \cdot \\ c_n \end{pmatrix} . \tag{12.2}$$

Then, the bra state $\langle a|$, which is a complex conjugate of the ket state $|a\rangle$, can be written as

$$\langle a| = \left(c_1^*, c_2^*, \ldots, c_n^* \right) . \tag{12.3}$$

12.2 Matrix Representation of Operators

Using the representation (12.1), we will try to write in a matrix form the relationship between two ket states $|a\rangle$ and $|b\rangle$ related through a linear operator $\hat{A}$ as

$$|b\rangle = \hat{A}|a\rangle . \tag{12.4}$$

Let

$$|a\rangle = \sum_n c_n |n\rangle ,$$

$$|b\rangle = \sum_m b_m |m\rangle . \tag{12.5}$$

Then

$$b_m = \langle m|b \rangle = \langle m|\hat{A}|a \rangle = \sum_n c_n \langle m|\hat{A}|n \rangle = \sum_n A_{mn} c_n, \quad (12.6)$$

where $A_{mn} = \langle m|\hat{A}|n \rangle$.

The right-hand side of Eq. (12.6) is the result of multiplication of a matrix composed of the elements A_{mn} and the column vector c_n:

$$\begin{pmatrix} b_1 \\ b_2 \\ \cdot \\ \cdot \\ \cdot \\ b_n \end{pmatrix} = \begin{pmatrix} A_{11} & A_{12} & \ldots & A_{1n} \\ A_{21} & A_{22} & \ldots & A_{2n} \\ \cdot & & & \\ \cdot & & & \\ \cdot & & & \\ A_{n1} & A_{n2} & \ldots & A_{nn} \end{pmatrix} \begin{pmatrix} c_1 \\ c_2 \\ \cdot \\ \cdot \\ \cdot \\ c_n \end{pmatrix}. \quad (12.7)$$

Thus, the scalar product $(\Psi_m, \hat{A}\Psi_n)$, or $\langle m|\hat{A}|n \rangle$ represents a matrix element of the operator $\hat{A}$ in the orthonormal basis $|n \rangle$. The matrix representing the operator is a square matrix of dimension $n \times n$.

The above procedure is very general and is applicable to any operator and an arbitrary orthonormal basis.

The following example will illustrate the introduced procedure of finding the matrix representation of a given operator in a given orthonormal basis.

Worked Example

Find the matrix representation of the operator $\hat{A} = d/dx$ in the basis of two orthonormal states $\Psi_1 = a \sin(nx)$ and $\Psi_2 = a \cos(nx)$, where $a = 1/\sqrt{\pi}$ and $x \in \langle -\pi, \pi \rangle$.

Solution

Since

$$\hat{A}\Psi_1 = \frac{d}{dx} a \sin(nx) = n\Psi_2,$$

$$\hat{A}\Psi_2 = \frac{d}{dx} a \cos(nx) = -n\Psi_1, \quad (12.8)$$

we find that the matrix elements of the operator $\hat{A}$ in the orthonormal basis Ψ_1, Ψ_2 are

$$\left(\Psi_1, \hat{A}\Psi_1\right) = n\left(\Psi_1, \Psi_2\right) = 0\,,$$
$$\left(\Psi_1, \hat{A}\Psi_2\right) = -n\left(\Psi_1, \Psi_1\right) = -n\,,$$
$$\left(\Psi_2, \hat{A}\Psi_2\right) = -n\left(\Psi_2, \Psi_1\right) = 0\,,$$
$$\left(\Psi_2, \hat{A}\Psi_1\right) = n\left(\Psi_2, \Psi_2\right) = n\,. \tag{12.9}$$

Hence, the matrix representation of the operator $\hat{A} = d/dx$ in the basis of two orthonormal states $\Psi_1 = a\sin(nx)$ and $\Psi_2 = a\cos(nx)$ is

$$\hat{A} = \begin{pmatrix} 0 & -n \\ n & 0 \end{pmatrix}\,. \tag{12.10}$$

We can solve this problem in a more elegant way using the Dirac notation.

Denote

$$|1\rangle = a\sin(nx)\,, \qquad |2\rangle = a\cos(nx)\,. \tag{12.11}$$

Since

$$\hat{A}|1\rangle = n|2\rangle\,,$$
$$\hat{A}|2\rangle = -n|1\rangle\,, \tag{12.12}$$

the operator $\hat{A}$ written in the basis $|1\rangle$, $|2\rangle$, has the form

$$\hat{A} = n\left(|2\rangle\langle 1| - |1\rangle\langle 2|\right)\,. \tag{12.13}$$

Hence

$$\langle 1|\hat{A}|1\rangle = 0\,, \qquad \langle 2|\hat{A}|2\rangle = 0\,,$$
$$\langle 1|\hat{A}|2\rangle = -n\,, \qquad \langle 2|\hat{A}|1\rangle = n\,. \tag{12.14}$$

Note that the operator $\hat{A}$ is not Hermitian

$$\hat{A}^\dagger = n\left(|1\rangle\langle 2| - |2\rangle\langle 1|\right) = -\hat{A}\,, \tag{12.15}$$

and, therefore, we can further conclude that the states $|1\rangle$, $|2\rangle$ are not the eigenfunctions of $\hat{A}$.

12.3 Matrix Representation of Eigenvalue Equations

We have seen that if we know the explicit forms of an operator $\hat{A}$ and a ket vector $|a\rangle$, we can easily determine whether the vector $|a\rangle$ is an eigenvector of the operator $\hat{A}$. Namely, the ket vector $|a\rangle$ is an eigenvector of a linear operator $\hat{A}$ if the ket vector $\hat{A}|a\rangle$ is a constant α times $|a\rangle$, i.e.,

$$\hat{A}|a\rangle = \alpha|a\rangle . \tag{12.16}$$

The complex constant α is called the eigenvalue and $|a\rangle$ is the eigenvector corresponding to the eigenvalue α.

In many practical situations, however, we know the explicit form of an operator $\hat{A}$, but we do not know explicit forms of the eigenvalues and eigenvectors of this operator. Then one can ask: Can we find the explicit forms of the eigenvalues and eigenvectors of the operator $\hat{A}$?

We *do* find the explicit forms of the eigenvalues and eigenvectors. We use the fact that an arbitrary vector $|a\rangle$ can always be expressed in terms of a linear superposition of arbitrary, but known, orthonormal vectors $|n\rangle$. Since $|a\rangle = \sum c_n|n\rangle$, we can write the eigenvalue equation as

$$\hat{A}|a\rangle = \sum_n c_n \hat{A}|n\rangle = \alpha \sum_m c_m|m\rangle . \tag{12.17}$$

Using the completeness relation to the left-hand side of Eq. (12.17), we get

$$\sum_n \sum_m c_n|m\rangle\langle m|\hat{A}|n\rangle = \alpha \sum_m c_m|m\rangle , \tag{12.18}$$

which can be written as

$$\sum_m \left(\sum_n c_n\langle m|\hat{A}|n\rangle \right) |m\rangle = \sum_m (\alpha c_m)|m\rangle . \tag{12.19}$$

Hence

$$\sum_n c_n\langle m|\hat{A}|n\rangle = \alpha c_m , \tag{12.20}$$

or

$$\sum_n c_n A_{mn} = \alpha c_m . \tag{12.21}$$

The left-hand side of Eq. (12.21) is the product of a column vector composed of the elements c_n and a matrix composed of the elements A_{mn}. Thus, we can write Eq. (12.21) in the matrix form as

$$\begin{pmatrix} A_{11} & A_{12} & \ldots & A_{1n} \\ A_{21} & A_{22} & \ldots & A_{2n} \\ & \cdot & & \\ & \cdot & & \\ & \cdot & & \\ A_{n1} & A_{n2} & \ldots & A_{nn} \end{pmatrix} \begin{pmatrix} c_1 \\ c_2 \\ \cdot \\ \cdot \\ \cdot \\ c_n \end{pmatrix} = \alpha \begin{pmatrix} c_1 \\ c_2 \\ \cdot \\ \cdot \\ \cdot \\ c_n \end{pmatrix}. \qquad (12.22)$$

This is a matrix eigenvalue equation.

The following conclusions arise from the matrix eigenvalue equation:

1. When the matrix $\bar{A}_{mn}$ is diagonal, i.e., $A_{mn} = 0$ for $m \neq n$, the orthonormal states $|n\rangle$ are the eigenstates of the operator $\hat{A}$ with eigenvalues $\alpha_n = A_{nn}$.
2. If the matrix $\bar{A}_{mn}$ is not diagonal, then we can find the eigenvalues and eigenvectors of $\hat{A}$ diagonalizing the matrix $\bar{A}_{mn}$. The eigenvalues are obtained from the characteristic equation

$$\begin{vmatrix} A_{11} - \alpha & A_{12} & \ldots & A_{1n} \\ A_{21} & A_{22} - \alpha & \ldots & A_{2n} \\ & \cdot & & \\ & \cdot & & \\ & \cdot & & \\ A_{n1} & A_{n2} & \ldots & A_{nn} - \alpha \end{vmatrix} = 0. \qquad (12.23)$$

This is of the form of a polynomial equation of degree n and shows that n eigenvalues can be found from the roots of the polynomial. For each eigenvalue α_i found by solving the characteristic equation, the corresponding eigenvector is found by substituting α_i into the matrix equation.

The following example will help to understand the procedure of finding the eigenvalues and eigenvectors of a given operator.

Worked Example

Consider the example from Section 12.2. In the matrix representation, the operator $\hat{A} = d/dx$ has the form given in Eq. (12.10). Since the matrix is *not* diagonal, the orthonormal states Ψ_1 and Ψ_2 are not eigenstates of the operator $\hat{A}$. We can find the eigenstates of $\hat{A}$ in terms of linear superpositions of the states Ψ_1 and Ψ_2, simply by the diagonalization of the matrix (12.10).

We start from the eigenvalue equation, which is of the form

$$\begin{pmatrix} 0 & -n \\ n & 0 \end{pmatrix} \begin{pmatrix} c_1 \\ c_2 \end{pmatrix} = \alpha \begin{pmatrix} c_1 \\ c_2 \end{pmatrix}. \tag{12.24}$$

First, we solve the characteristic equation

$$\begin{vmatrix} -\alpha & -n \\ n & -\alpha \end{vmatrix} = 0, \tag{12.25}$$

from which we find two eigenvalues $\alpha_1 = +in$ and $\alpha_2 = -in$.

For $\alpha_1 = in$, the eigenvalue equation takes the form

$$\begin{pmatrix} 0 & -n \\ n & 0 \end{pmatrix} \begin{pmatrix} c_1 \\ c_2 \end{pmatrix} = in \begin{pmatrix} c_1 \\ c_2 \end{pmatrix}, \tag{12.26}$$

from which, we find that

$$-nc_2 = inc_1 \quad \Rightarrow \quad c_1 = ic_2. \tag{12.27}$$

Hence, the eigenfunction corresponding to the eigenvalue α_1 is of the form

$$\Psi_{\alpha_1} = \begin{pmatrix} ic_2 \\ c_2 \end{pmatrix} = c_2 \begin{pmatrix} i \\ 1 \end{pmatrix}. \tag{12.28}$$

From the normalization of Ψ_{α_1}, we get

$$1 = \left(\Psi_{\alpha_1}, \Psi_{\alpha_1} \right) = |c_2|^2 \, (-i, 1) \begin{pmatrix} i \\ 1 \end{pmatrix} = 2 \, |c_2|^2. \tag{12.29}$$

Thus, the normalized eigenfunction corresponding to the eigenvalue α_1 is given by

$$\Psi_{\alpha_1} = \frac{1}{\sqrt{2}} \begin{pmatrix} i \\ 1 \end{pmatrix}, \tag{12.30}$$

which may be written as

$$\Psi_{\alpha_1} = \frac{1}{\sqrt{2\pi}} \, [i \sin(nx) + \cos(nx)] = \frac{1}{\sqrt{2\pi}} e^{inx}. \tag{12.31}$$

Similarly, we can easily show that the normalized eigenfunction corresponding to the eigenvalue α_2 is of the form

$$\Psi_{\alpha_2} = \frac{1}{\sqrt{2\pi}} [-i \sin(nx) + \cos(nx)] = \frac{1}{\sqrt{2\pi}} e^{-inx} . \quad (12.32)$$

In the Dirac notation, the normalized eigenvectors can be written in a compact form as

$$|\alpha_1\rangle = \frac{1}{\sqrt{2}} (i|1\rangle + |2\rangle) ,$$

$$|\alpha_2\rangle = \frac{1}{\sqrt{2}} (-i|1\rangle + |2\rangle) . \quad (12.33)$$

The physical interpretation of the superposition states (12.33) is as follows: The eigenfunctions $|\alpha_1\rangle$ and $|\alpha_2\rangle$ in the form of the linear superpositions tell us that, e.g., with the probability $1/2$, the system described by the operator $\hat{A}$ is in the state $|1\rangle$ or in the state $|2\rangle$.

In summary of this chapter, we have learned that

(1) In quantum physics, an arbitrary wave function may be represented by a normalized column vector of expansion coefficients in the basis of orthonormal states.
(2) In an orthonormal basis, an arbitrary operator $\hat{A}$ may be represented by a matrix, whose elements A_{mn} are given by scalar products $(\Psi_m, \hat{A}\Psi_n)$.
(3) Using an orthonormal basis, an eigenvalue equation of an arbitrary operator may be written in a matrix form. In this case, the problem of finding eigenvalues and eigenvectors of the operator reduces to a simple problem of diagonalization of the matrix.

Tutorial Problems

Problem 12.1 *Eigenvalues and eigenvectors of Hermitian operators.*

(a) Consider two Hermitian operators $\hat{A}$ and $\hat{B}$ that have the same complete set of eigenfunctions ϕ_n. Show that the operators commute.

(b) Suppose two Hermitian operators have the matrix representation:

$$\hat{A} = \begin{bmatrix} a & 0 & 0 \\ 0 & -a & 0 \\ 0 & 0 & -a \end{bmatrix}, \quad \hat{B} = \begin{bmatrix} b & 0 & 0 \\ 0 & 0 & ib \\ 0 & -ib & 0 \end{bmatrix},$$

where a and b are real numbers.

(i) Calculate the eigenvalues of $\hat{A}$ and $\hat{B}$.
(ii) Show that $\hat{A}$ and $\hat{B}$ commute.
(iii) Determine a complete set of common eigenfunctions.

Problem 12.2 *The Rabi problem that illustrates what are the energy states of an atom driven by an external coherent (laser) field.*

A laser field of frequency ω_L drives a transition in an atom between two atomic energy states $|1\rangle$ and $|2\rangle$. The states are separated by the frequency ω_0. The Hamiltonian of the system in the bases of the atomic states is given by the matrix

$$\hat{H} = \hbar \begin{pmatrix} -\frac{1}{2}\Delta & \Omega \\ \Omega & \frac{1}{2}\Delta \end{pmatrix},$$

where $\Delta = \omega_L - \omega_0$ is the detuning of the laser frequency from the atomic transition frequency, and Ω is the Rabi frequency that describes the strength of the laser field acting on the atom.

Find the energies and energy states of the system, the so-called dressed states, which are, respectively, eigenvalues and eigenvectors of the Hamiltonian $\hat{H}$.

Chapter 13

Spin Operators and Pauli Matrices

It is well known from atomic spectroscopy and in particular from the Stern–Gerlach experiment that electrons have a nonzero angular momentum (spin) and that there exist only two kinds of electrons with spin $\hbar/2$ or $-\hbar/2$. An external field or force cannot change the value of the electron spin; they can only change the orientation of the spin in space. This means that the projection of the spin on any of the coordinate axis is always equal to $\pm\hbar/2$. In this chapter, we will derive the Pauli matrices, the matrix representation of the spin. We will also demonstrate how the two kinds of spin can be represented in terms of spin up and spin down operators.

For convenience, the spin of the electron can be written as

$$\vec{S} = \frac{1}{2}\hbar\vec{\sigma},\qquad(13.1)$$

where from now the new operator $\vec{\sigma}$ will be called the spin operator. From the above relation, we see that the eigenvalues of the operator $\vec{\sigma}$ are $+1$ and -1.

Note that the magnitude of $\vec{\sigma}$, i.e., $\sigma^2 = \vec{\sigma}\cdot\vec{\sigma} = 1$. Thus, for any direction in the cartesian coordinates

$$\sigma_x^2 = \sigma_y^2 = \sigma_z^2 = 1.\qquad(13.2)$$

Since the spin is an angular momentum of the electron, it should satisfy the commutation relations for the components of the angular

Quantum Physics for Beginners
Zbigniew Ficek
Copyright © 2016 Pan Stanford Publishing Pte. Ltd.
ISBN 978-981-4669-38-2 (Hardcover), 978-981-4669-39-9 (eBook)
www.panstanford.com

momentum

$$[\sigma_i, \sigma_j] = 2ie_{ijk}\sigma_k. \quad i, j, k = x, y, z. \tag{13.3}$$

Using Eqs. (13.2) and (13.3), we find that

$$2i(\sigma_y\sigma_z + \sigma_z\sigma_y) = \sigma_y(2i\sigma_z) + (2i\sigma_z)\sigma_y$$
$$= \sigma_y(\sigma_x\sigma_y - \sigma_y\sigma_x) + (\sigma_x\sigma_y - \sigma_y\sigma_x)\sigma_y$$
$$= \sigma_y\sigma_x\sigma_y - \sigma_x + \sigma_x - \sigma_y\sigma_x\sigma_y = 0. \tag{13.4}$$

In the same way, we can show that

$$2i(\sigma_x\sigma_y + \sigma_y\sigma_x) = 0,$$
$$2i(\sigma_z\sigma_x + \sigma_x\sigma_z) = 0. \tag{13.5}$$

These results show that the components of the spin anti-commute:

$$[\sigma_i, \sigma_j]_+ = 0, \quad i \neq j, \quad i, j = x, y, z. \tag{13.6}$$

13.1 Matrix Representation of the Spin Operators: Pauli Matrices

We wish to find the matrix representation of the spin operators. To solve this problem, we use the conditions imposed on the spin components, the commutation and anti-commutation relations, Eqs. (13.3) and (13.6).

Note first that the trace of the matrices is zero.

$$\text{Tr}\,\sigma_i = 0, \quad i = x, y, z. \tag{13.7}$$

Proof: Using Eqs. (13.2) and (13.6), we find

$$\text{Tr}\,\sigma_i = \text{Tr}\,\sigma_i\sigma_k\sigma_k = -\text{Tr}\,\sigma_k\sigma_i\sigma_k = -\text{Tr}\,\sigma_i\sigma_k\sigma_k = -\text{Tr}\,\sigma_i, \tag{13.8}$$

which is satisfied only when $\text{Tr}\,\sigma_i = 0$. In the proof, we have used the cycling property of the trace that $\text{Tr}\,ABC = \text{Tr}\,BCA$.

Our problem may now be stated as follows: Obtain the matrix representation of the spin components such that in the basis of the spin states, spin "up" $|2\rangle$ and spin "down" $|1\rangle$, the component σ_z is represented by a diagonal matrix

$$\sigma_z = \begin{pmatrix} 1 & 0 \\ 0 & -1 \end{pmatrix}. \tag{13.9}$$

Then, what form are the matrices of the other components, σ_x and σ_y? To answer this question, consider the operator σ_x, which written in the basis $|2\rangle$ and $|1\rangle$ has a general matrix form

$$\sigma_x = \begin{pmatrix} a_{22} & a_{21} \\ a_{12} & a_{11} \end{pmatrix}, \tag{13.10}$$

where a_{ij} are unknown matrix elements that we have to determine.

In order to determine the matrix elements, we use the anti-commutation relations (13.6), from which we find that

$$0 = \sigma_z\sigma_x + \sigma_x\sigma_z = 2 \begin{pmatrix} a_{22} & 0 \\ 0 & -a_{11} \end{pmatrix}, \tag{13.11}$$

which means that $a_{11} = a_{22} = 0$.

Since the spin operator is Hermitian, as it represents an observable (physical quantity), so then the components of the spin are also Hermitian. Thus, we have that

$$\sigma_x = \begin{pmatrix} 0 & a_{21} \\ a_{12} & 0 \end{pmatrix} = \begin{pmatrix} 0 & a_{21} \\ a_{21}^* & 0 \end{pmatrix}. \tag{13.12}$$

If we take the square of σ_x, we get

$$\sigma_x^2 = \begin{pmatrix} a_{21}a_{21}^* & 0 \\ 0 & a_{21}^*a_{21} \end{pmatrix} = \begin{pmatrix} |a_{21}|^2 & 0 \\ 0 & |a_{21}|^2 \end{pmatrix} = 1. \tag{13.13}$$

Hence $|a_{21}|^2 = 1$, so that $a_{21} = e^{i\alpha}$, where α is an arbitrary real number.

Thus, the component σ_x may be written in the matrix form as

$$\sigma_x = \begin{pmatrix} 0 & e^{i\alpha} \\ e^{-i\alpha} & 0 \end{pmatrix}. \tag{13.14}$$

Similarly, for σ_y, we may show that

$$\sigma_y = \begin{pmatrix} 0 & e^{i\beta} \\ e^{-i\beta} & 0 \end{pmatrix}. \tag{13.15}$$

In order to determine α and β, we apply the above matrices to the anti-commutation relation, and find

$$0 = \sigma_x\sigma_y + \sigma_y\sigma_x = \begin{pmatrix} e^{i(\alpha-\beta)} + e^{-i(\alpha-\beta)} & 0 \\ 0 & e^{-i(\alpha-\beta)} + e^{i(\alpha-\beta)} \end{pmatrix}$$
$$= 2 \begin{pmatrix} \cos(\alpha - \beta) & 0 \\ 0 & \cos(\alpha - \beta) \end{pmatrix}. \tag{13.16}$$

This means that
$$\cos(\alpha - \beta) = 0, \tag{13.17}$$
which is met if
$$\alpha - \beta = (2n + 1)\frac{\pi}{2}, \quad n = 0, 1, 2, \ldots \tag{13.18}$$

If we choose the simplest values for n and α: $n = 0$ and $\alpha = 0$, we then have $\beta = -\pi/2$. Hence, the matrices σ_x and σ_y take the form
$$\sigma_x = \begin{pmatrix} 0 & 1 \\ 1 & 0 \end{pmatrix}, \quad \sigma_y = \begin{pmatrix} 0 & -i \\ i & 0 \end{pmatrix}. \tag{13.19}$$

Thus, the components of the spin can be represented by matrices
$$\sigma_x = \begin{pmatrix} 0 & 1 \\ 1 & 0 \end{pmatrix}, \quad \sigma_y = \begin{pmatrix} 0 & -i \\ i & 0 \end{pmatrix}, \quad \sigma_z = \begin{pmatrix} 1 & 0 \\ 0 & -1 \end{pmatrix}, \tag{13.20}$$
called the Pauli matrices.

The Pauli matrices have many interesting properties. For example, the reader may immediately notice from Eq. (13.20) that the trace of the matrices is zero. The detailed study of the other properties of the Pauli matrices is left for the reader as a tutorial problem (see Tutorial Problem 13.4).

13.2 Spin "Up" and Spin "Down" Operators

The operators $\hat{\sigma}_x$, $\hat{\sigma}_y$, and $\hat{\sigma}_z$ representing the components of the electron spin can be written in terms of the spin raising and spin lowering operators σ^+ and σ^- as
$$\hat{\sigma}_x = \hat{\sigma}^+ + \hat{\sigma}^-, \tag{13.21}$$
$$\hat{\sigma}_y = \left(\hat{\sigma}^+ - \hat{\sigma}^-\right)/i, \tag{13.22}$$
$$\hat{\sigma}_z = \hat{\sigma}^+\hat{\sigma}^- - \hat{\sigma}^-\hat{\sigma}^+. \tag{13.23}$$

Let $|1\rangle$ and $|2\rangle$ be the two eigenstates of the electron spin. The raising and lowering operators satisfy the following relations:
$$\hat{\sigma}^+|1\rangle = |2\rangle, \quad \hat{\sigma}^-|1\rangle = 0, \tag{13.24}$$
$$\hat{\sigma}^+|2\rangle = 0, \quad \hat{\sigma}^-|2\rangle = |1\rangle. \tag{13.25}$$

The relations explicitly illustrate where the meaning of the spin raising and lowering comes from. The operator σ^+ turns the "down" spin into the "up" spin. In other words, it raises the spin. The operator σ^- acts in the opposite direction. It turns the "up" spin into the "down" spin, that it lowers the spin.

Tutorial Problems

Problem 13.1 Calculate the square $(\vec{A} \cdot \vec{S})^2$ of the scalar product of an arbitrary vector $\vec{A}$ and of the spin vector $\vec{S} = S_x \vec{i} + S_y \vec{j} + S_z \vec{k}$.

Problem 13.2 Calculate the squares of the spin components, σ_x^2, σ_y^2, and σ_z^2, and verify if the squares of the spin components can be simultaneously measured with the same precision.

Problem 13.3 *Matrix representation of the spin operators*
The operators $\hat{\sigma}_x$, $\hat{\sigma}_y$, and $\hat{\sigma}_z$ representing the components of the electron spin can be written in terms of the spin raising and spin lowering operators σ^+ and σ^- as

$$\hat{\sigma}_x = \hat{\sigma}^+ + \hat{\sigma}^- ,$$
$$\hat{\sigma}_y = \left(\hat{\sigma}^+ - \hat{\sigma}^- \right) / i ,$$
$$\hat{\sigma}_z = \hat{\sigma}^+ \hat{\sigma}^- - \hat{\sigma}^- \hat{\sigma}^+ .$$

Let $|1\rangle$ and $|2\rangle$ be the two eigenstates of the electron spin with the eigenvalues $-\hbar/2$ and $+\hbar/2$, respectively, as determined in the Stern–Gerlach experiment. The raising and lowering operators satisfy the following relations:

$$\hat{\sigma}^+ |1\rangle = |2\rangle , \qquad \hat{\sigma}^- |1\rangle = 0 ,$$
$$\hat{\sigma}^+ |2\rangle = 0 , \qquad \hat{\sigma}^- |2\rangle = |1\rangle .$$

Using these relations, find the matrix representations (the Pauli matrices) of the operators $\hat{\sigma}_x$, $\hat{\sigma}_y$, and $\hat{\sigma}_z$ in the basis of the states $|1\rangle$ and $|2\rangle$.

Problem 13.4 *Properties of the Pauli matrices*
Consider the Pauli matrices representing the spin operators $\hat{\sigma}_x$, $\hat{\sigma}_y$, and $\hat{\sigma}_z$ in the basis of the states $|1\rangle$ and $|2\rangle$.

(a) Prove that the operators $\hat{\sigma}_x$, $\hat{\sigma}_y$, $\hat{\sigma}_z$ are Hermitian. This result is what the readers could expect as the operators represent a physical (measurable) quantity, the electron spin.
(b) Show that the operators $\hat{\sigma}_x$, $\hat{\sigma}_y$, $\hat{\sigma}_z$ each has eigenvalues $+1$, -1. Determine the normalized eigenvectors of each. Are $|1\rangle$ and $|2\rangle$ the eigenvectors of any of the matrices?

(c) Show that the operators $\hat{\sigma}_x, \hat{\sigma}_y, \hat{\sigma}_z$ obey the commutation relations

$$\left[\hat{\sigma}_x, \hat{\sigma}_y\right] = 2i\hat{\sigma}_z ,$$
$$\left[\hat{\sigma}_z, \hat{\sigma}_x\right] = 2i\hat{\sigma}_y ,$$
$$\left[\hat{\sigma}_y, \hat{\sigma}_z\right] = 2i\hat{\sigma}_x .$$

If you recall the Heisenberg uncertainty relation, you will conclude immediately that these commutation relations show that the three components of the spin cannot be measured simultaneously with the same precision.

(d) Calculate anti-commutators $\left[\hat{\sigma}_x, \hat{\sigma}_y\right]_+, \left[\hat{\sigma}_z, \hat{\sigma}_x\right]_+, \left[\hat{\sigma}_y, \hat{\sigma}_z\right]_+.$

(e) Show that $\hat{\sigma}_x^2 = \hat{\sigma}_y^2 = \hat{\sigma}_z^2 = \hat{1}.$ This result is a confirmation of the conservation of the total spin of the system that the magnitude of the total spin vector is constant.

(f) Write the operators $\hat{\sigma}_x, \hat{\sigma}_y,$ and $\hat{\sigma}_z$ in terms of the projection operators $\hat{P}_{ij} = |i\rangle\langle j|, (i, j = 1, 2).$

Chapter 14

Quantum Dynamics and Pictures

We have already learned that to find the evolution of a physical system, we have to specify how state vectors and operators of a given system evolve in time. In this chapter, we will demonstrate how one can find the time evolution of the operators from the knowledge of the time evolution of the state vector, and vice versa. We introduce a unitary operator and discuss the concept of a unitary transformation to show how one could transfer the time dependence from the state vectors to the operators. The possibility that either the state vectors or the operators can depend explicitly on time will lead us to introduce the Schrödinger, Heisenberg, and interaction pictures. We will discuss the fundamental differences between these pictures. Finally, we discuss the Ehrenfest theorem, which shows under which conditions quantum mechanics predicts the same results for measured physical quantities as classical physics.

14.1 Unitary Time-Evolution Operator

Consider the time-dependent Schrödinger equation

$$i\hbar\frac{\partial}{\partial t}|\Psi(\vec{r}, t)\rangle = \hat{H}_t|\Psi(\vec{r}, t)\rangle, \tag{14.1}$$

Quantum Physics for Beginners
Zbigniew Ficek
Copyright © 2016 Pan Stanford Publishing Pte. Ltd.
ISBN 978-981-4669-38-2 (Hardcover), 978-981-4669-39-9 (eBook)
www.panstanford.com

where $\hat{H}_t \equiv \hat{H}(\vec{r}, t)$. If the Hamiltonian is independent of time, i.e., $\hat{H}_t = \hat{H}$, the solution to the Schrödinger equation (14.1) is of the form

$$|\Psi(\vec{r}, t)\rangle = e^{-i\frac{\hat{H}}{\hbar}(t-t_0)}|\Psi(\vec{r}, t_0)\rangle \equiv \hat{U}(t, t_0)|\Psi(\vec{r}, t_0)\rangle, \quad (14.2)$$

where

$$\hat{U}(t, t_0) = e^{-i\frac{\hat{H}}{\hbar}(t-t_0)}, \quad (14.3)$$

is a unitary time-evolution operator.

A unitary operator should satisfy the following property: $\hat{U}(t, t_0)\hat{U}^{\dagger}(t, t_0) = 1$, where $\hat{U}^{\dagger}(t, t_0)$ is the adjoint operator of $\hat{U}(t, t_0)$. For $\hat{U}(t, t_0)$ of the form (14.3), we use the Taylor expansion and find

$$\hat{U}(t, t_0)\hat{U}^{\dagger}(t, t_0) = e^{-i\frac{\hat{H}}{\hbar}(t-t_0)}e^{i\frac{\hat{H}}{\hbar}(t-t_0)}$$

$$= \left[1 - i\frac{\hat{H}}{\hbar}(t-t_0) + \frac{1}{2}\left(-i\frac{\hat{H}}{\hbar}(t-t_0)\right)^2 + \ldots\right]$$

$$\times \left[1 + i\frac{\hat{H}}{\hbar}(t-t_0) + \frac{1}{2}\left(i\frac{\hat{H}}{\hbar}(t-t_0)\right)^2 + \ldots\right]$$

$$= 1 + i\frac{\hat{H}}{\hbar}(t-t_0) - i\frac{\hat{H}}{\hbar}(t-t_0) + \ldots = 1. \quad (14.4)$$

In a similar way, we can show that $\hat{U}^{\dagger}(t, t_0)\hat{U}(t, t_0) = 1$. Obviously, $\hat{U}(t_0, t_0) = 1$.

In addition, $\hat{U}(t, t_0)$ satisfies the transitive feature

$$\hat{U}(t, t_1)\hat{U}(t_1, t_0) = \hat{U}(t, t_0). \quad (14.5)$$

Proof: According to the solution (14.2), we can write for an arbitrary $t > t_1 > t_0$:

$$|\Psi(\vec{r}, t)\rangle = \hat{U}(t, t_1)|\Psi(\vec{r}, t_1)\rangle = \hat{U}(t, t_1)\hat{U}(t_1, t_0)|\Psi(\vec{r}, t_0)\rangle. \quad (14.6)$$

Since $|\Psi(\vec{r}, t)\rangle = \hat{U}(t, t_0)|\Psi(\vec{r}, t_0)\rangle$, we have that $\hat{U}(t, t_1)\hat{U}(t_1, t_0) = \hat{U}(t, t_0)$.

The state vector (14.2) is given in terms of the initial state $|\Psi(\vec{r}, t_0)\rangle$. If we know the eigenstates $|\phi_n\rangle$ of the Hamiltonian $\hat{H}$, then we can use the completeness relation for the eigenstates, $\sum_n |\phi_n\rangle\langle\phi_n| = 1$, and write the state (14.2) in terms of $|\phi_n\rangle$ as

$$|\Psi(\vec{r}, t)\rangle = \hat{U}(t, t_0)|\Psi(\vec{r}, t_0)\rangle = \sum_n \hat{U}(t, t_0)|\phi_n\rangle\langle\phi_n|\Psi(\vec{r}, t_0)\rangle$$

$$= \sum_n e^{-i\frac{E_n}{\hbar}(t-t_0)}|\phi_n\rangle\langle\phi_n|\Psi(\vec{r}, t_0)\rangle = \sum_n c_n e^{-i\frac{E_n}{\hbar}(t-t_0)}|\phi_n\rangle,$$

$$\tag{14.7}$$

where $c_n = \langle\phi_n|\Psi(\vec{r}, t_0)\rangle$ is the probability amplitude that at time $t = 0$, the system was in the state $|\phi_n\rangle$.

14.2 Unitary Transformation of State Vectors

We can make an arbitrary unitary transformation of the state vector $|\Psi(\vec{r}, t)\rangle$ to a "new" state vector, which will depend on a different time or to a state vector independent of time. For example, if we make a transformation

$$\hat{U}^\dagger(t, t_0)|\Psi(\vec{r}, t\rangle = |\Psi(\vec{r}, t)\rangle_T, \tag{14.8}$$

the new "transformed" vector $|\Psi(\vec{r}, t)\rangle_T$ is independent of time. It is easy to see. If we use the result (14.7), we readily find

$$|\Psi(\vec{r}, t)\rangle_T = \hat{U}^\dagger(t, t_0)|\Psi(\vec{r}, t)\rangle = \sum_n c_n e^{-i\frac{E_n}{\hbar}(t-t_0)}\hat{U}^\dagger(t, t_0)|\phi_n\rangle$$

$$= \sum_n c_n e^{-i\frac{E_n}{\hbar}(t-t_0)} e^{i\frac{\hat{H}}{\hbar}(t-t_0)}|\phi_n\rangle$$

$$= \sum_n c_n e^{-i\frac{E_n}{\hbar}(t-t_0)} e^{i\frac{E_n}{\hbar}(t-t_0)}|\phi_n\rangle$$

$$= \sum_n c_n|\phi_n\rangle = |\Psi(\vec{r}, t_0)\rangle. \tag{14.9}$$

The transformed vector is equal to the initial vector $|\Psi(\vec{r}, t_0)\rangle$.

We can make unitary transformations not only of the state vectors but also the Schrödinger equation. In other words, if we know the Schrödinger equation for the state vector $|\Psi(\vec{r}, t)\rangle$, we can find the evolution of motion of the transformed state vector. To show this, we start from the time-dependent Schrödinger equation (14.1) and using the fact that $|\Psi(\vec{r}, t)\rangle = \hat{U}|\Psi(\vec{r}, t)\rangle_T$, we find

$$i\hbar\frac{\partial}{\partial t}|\Psi(\vec{r}, t)\rangle = i\hbar\frac{\partial}{\partial t}\left(\hat{U}|\Psi(\vec{r}, t)\rangle_T\right) = \hat{H}_t\hat{U}|\Psi(\vec{r}, t)\rangle_T. \tag{14.10}$$

Since

$$i\hbar \frac{\partial}{\partial t} \left(\hat{U} \, |\Psi(\vec{r}, t)\rangle_T \right) = i\hbar \left[\left(\frac{\partial}{\partial t} \hat{U} \right) |\Psi(\vec{r}, t)\rangle_T + \hat{U} \frac{\partial}{\partial t} |\Psi(\vec{r}, t)\rangle_T \right]$$

$$= i\hbar \left[-\frac{i}{\hbar} \hat{U} \, \hat{H} \, |\Psi(\vec{r}, t)\rangle_T + \hat{U} \frac{\partial}{\partial t} |\Psi(\vec{r}, t)\rangle_T \right]$$

$$= \hat{U} \, \hat{H} \, |\Psi(\vec{r}, t)\rangle_T + i\hbar \hat{U} \frac{\partial}{\partial t} |\Psi(\vec{r}, t)\rangle_T, \quad (14.11)$$

we get

$$\hat{U} \, \hat{H} \, |\Psi(\vec{r}, t)\rangle_T + i\hbar \hat{U} \frac{\partial}{\partial t} |\Psi(\vec{r}, t)\rangle_T = \hat{H}_t \hat{U} \, |\Psi(\vec{r}, t)\rangle_T. \quad (14.12)$$

Multiplying both sides from the left by $\hat{U}^\dagger$, we obtain

$$\hat{H} \, |\Psi(\vec{r}, t)\rangle_T + i\hbar \frac{\partial}{\partial t} |\Psi(\vec{r}, t)\rangle_T = \hat{U}^\dagger \hat{H}_t \hat{U} \, |\Psi(\vec{r}, t)\rangle_T, \quad (14.13)$$

which can be written as

$$i\hbar \frac{\partial}{\partial t} |\Psi(\vec{r}, t)\rangle_T = \left(\hat{U}^\dagger \hat{H}_t \hat{U} - \hat{H} \right) |\Psi(\vec{r}, t)\rangle_T = \hat{H}_T |\Psi(\vec{r}, t)\rangle_T,$$

$$(14.14)$$

where $\hat{H}_T = \hat{U}^\dagger \hat{H}_t \hat{U} - \hat{H}$. We see that the transformed state satisfies the Schrödinger equation with the effective (transformed) Hamiltonian $\hat{H}_T$.

14.3 Unitary Transformation of Operators

Let us now illustrate unitary transformations of an arbitrary operator $\hat{A}(t)$, which can depend explicitly on time.

Consider the mean value of the operator $\langle \hat{A}(t) \rangle = \langle \Psi(t) | \hat{A}(t) | \Psi(t) \rangle$ and a transformed operator $\hat{A}'(t) = \hat{U}^\dagger \hat{A}(t) \hat{U}$. We can make unitary transformations of mean values of the operator. For example, the mean value of the transformed operator is

$$_T \langle \Psi(t) | \hat{A}'(t) | \Psi(t) \rangle_T = \langle \Psi(t) | \hat{U} \, \hat{U}^\dagger \hat{A}(t) \hat{U} \, \hat{U}^\dagger | \Psi(t) \rangle = \langle \Psi(t) | \hat{A}(t) | \Psi(t) \rangle.$$

$$(14.15)$$

We see that the mean value of the transformed operator is the same as the mean value of $\hat{A}(t)$.

We can also make a transformation of the equation of motion for the operator $\hat{A}(t)$. Calculate a time derivative of $\hat{A}'(t)$ and find

$$\frac{d}{dt}\hat{A}'(t) = \left(\frac{\partial}{\partial t}\hat{U}^\dagger\right)\hat{A}\hat{U} + \hat{U}^\dagger\left(\frac{\partial}{\partial t}\hat{A}\right)\hat{U} + \hat{U}^\dagger\hat{A}\left(\frac{\partial}{\partial t}\hat{U}\right)$$

$$= \frac{i}{\hbar}\hat{U}^\dagger\hat{H}\hat{A}\hat{U} + \hat{U}^\dagger\left(\frac{\partial}{\partial t}\hat{A}\right)\hat{U} - \frac{i}{\hbar}\hat{U}^\dagger\hat{A}\hat{H}\hat{U}$$

$$= \frac{i}{\hbar}\hat{U}^\dagger\hat{H}\hat{U}\hat{U}^\dagger\hat{A}\hat{U} + \hat{U}^\dagger\left(\frac{\partial}{\partial t}\hat{A}\right)\hat{U} - \frac{i}{\hbar}\hat{U}^\dagger\hat{A}\hat{U}\hat{U}^\dagger\hat{H}\hat{U}$$

$$= \frac{i}{\hbar}\hat{H}'\hat{A}' + \hat{U}^\dagger\left(\frac{\partial}{\partial t}\hat{A}\right)\hat{U} - \frac{i}{\hbar}\hat{A}'\hat{H}'$$

$$= \hat{U}^\dagger\left(\frac{\partial}{\partial t}\hat{A}\right)\hat{U} + \frac{i}{\hbar}\left[\hat{H}', \hat{A}'\right], \tag{14.16}$$

where $\hat{H}' = \hat{U}^\dagger\hat{H}\hat{U}$. An important conclusion arises from Eq. (14.16). We see that the transformed operator $\hat{A}'(t)$ depends on time even if the operator $\hat{A}(t)$ does not depend explicitly on time $(\partial\hat{A}(t)/\partial t = 0)$, unless the commutator $[\hat{H}', \hat{A}'] = 0$, and then $\hat{A}'(t)$ is independent of time.

We have shown that the evolution of the transformed state $|\Psi(t)\rangle_T$ depends on whether the Hamiltonian of a given system depends explicitly on time or not. There could be situations where the state vector of a given system depends on time but the operators of the system are independent of time, and vice versa. Thus, we can distinguish different cases, which are called "pictures." Depending on whether states or operators depend explicitly on time, we distinguish three pictures: the Schrödinger, Heisenberg, and interaction pictures.

14.4 Schrödinger Picture

Consider a system whose Hamiltonian $\hat{H}_t$ is independent of time, i.e., $\hat{H}_t = \hat{H}$. In this case, only the state vector depends on time. The dependence arises from the fact that the wave function satisfies the time-dependent Schrödinger equation

$$i\hbar\frac{\partial}{\partial t}|\Psi(\vec{r}, t)\rangle = \hat{H}|\Psi(\vec{r}, t)\rangle, \tag{14.17}$$

whose solution is of the form

$$|\Psi_s(\vec{r}, t)\rangle = e^{-i\frac{\hat{H}}{\hbar}(t-t_0)}|\Psi(\vec{r}, t_0)\rangle. \tag{14.18}$$

A situation where the operators are independent of time but the state vectors depend explicitly on time is called the *Schrödinger picture*. In Eq. (14.18), the subscript "*s*" indicates that the evolution of the system is described in the Schrödinger picture that the wave function depends explicitly on time and the operator $\hat{H}$ is independent of time.

14.5 Heisenberg Picture

When the state vectors of a given system are independent of time, but the operators depend on time, we call it the *Heisenberg picture*. For example, if in Eq. (14.14), the Hamiltonian $\hat{H}_t$ is independent of time, i.e., $\hat{H}_t = \hat{H}$ then $\hat{H}_T = 0$, which results in

$$i\hbar \frac{\partial}{\partial t} |\Psi(\vec{r}, t)\rangle_T = 0. \tag{14.19}$$

Thus, the transformed state vector is independent of time.

Consider now the evolution of an operator. Let us introduce a time-independent operator $\hat{A}$ and its unitary transformation $\hat{A}'(t) = \hat{U}^\dagger \hat{A} \hat{U}$. We have already seen, Eq. (14.16), that the transformed operator depends explicitly on time as it satisfies the Heisenberg equation of motion

$$\frac{d}{dt} \hat{A}'(t) = \frac{i}{\hbar} \left[\hat{H}', \hat{A}' \right]. \tag{14.20}$$

The state vector $|\Psi(\vec{r}, t)\rangle_T$ and the operator $\hat{A}'(t)$ are called, respectively, the state vector and the operator in the Heisenberg picture $(|\Psi(\vec{r}, t)\rangle_T \equiv |\Psi_H(\vec{r}, t)\rangle, \hat{A}'(t) = \hat{A}_H(t))$.

In summary, in the Heisenberg picture, the state vectors are independent of time, whereas operators depend explicitly on time and satisfy the Heisenberg equation of motion

$$\frac{d}{dt} \hat{A}_H(t) = \frac{i}{\hbar} \left[\hat{H}_H(t), \hat{A}_H(t) \right]. \tag{14.21}$$

Note that the transition from the Schrödinger to the Heisenberg pictures is in fact the transfer of the time dependence from the state vectors to the operators. It should be pointed out that the transfer of the time dependence does not change measured quantities such as averages. For example, the average value of an operator $\hat{A}$ in the Schrödinger and Heisenberg pictures has the same value

$$\langle \Psi_s(\vec{r}, t)|\hat{A}|\Psi_s(\vec{r}, t)\rangle = {}_T\langle \Psi(\vec{r}, t)|\hat{U}^\dagger \hat{A}\hat{U}|\Psi(\vec{r}, t)\rangle_T$$
$$= \langle \Psi_H(\vec{r}, t)|\hat{A}_H(t)|\Psi_H(\vec{r}, t)\rangle. \quad (14.22)$$

Worked Example

Find the Heisenberg equations of motion for the position $\hat{x}$ and momentum $\hat{p}_x$ operators of a single particle of mass m moving in one dimension in a potential energy $\hat{V}(x)$.

Solution

The Hamiltonian of the particle moving in one dimension is given by

$$\hat{H} = \frac{1}{2m}\hat{p}_x^2 + \hat{V}(x), \quad (14.23)$$

and we have the well-known commutation relations

$$[\hat{x}, \hat{p}_x] = i\hbar, \quad [\hat{V}(x), \hat{x}] = 0, \quad [\hat{V}(x), \hat{p}_x] = i\hbar\frac{\partial \hat{V}(x)}{\partial x}. \quad (14.24)$$

Assuming that the operators $\hat{x}$ and $\hat{p}_x$ do not depend explicitly on time, we have the following equation of motion for the position operator:

$$\frac{d\hat{x}}{dt} = \frac{i}{\hbar}\left[\hat{H}, \hat{x}\right] = \frac{i}{2m\hbar}\left[\hat{p}_x^2, \hat{x}\right] = -\frac{i}{2m\hbar}\left[\hat{x}, \hat{p}_x^2\right]$$
$$= -\frac{i}{2m\hbar}\{[\hat{x}, \hat{p}_x]\hat{p}_x + \hat{p}_x[\hat{x}, \hat{p}_x]\} = \frac{1}{m}\hat{p}_x. \quad (14.25)$$

Hence

$$\hat{p}_x = m\frac{d\hat{x}}{dt}. \quad (14.26)$$

Notice that the Heisenberg equation of motion for the position operator is analogous to the classical equation

$$p_x = m\frac{dx}{dt}. \quad (14.27)$$

The Heisenberg equation of motion for the momentum operator is found in the following way:

$$\frac{d\hat{p}_x}{dt} = \frac{i}{\hbar}\left[\hat{H}, \hat{p}_x\right] = \frac{i}{\hbar}\left[\hat{V}(x), \hat{p}_x\right] = -\frac{\partial\hat{V}(x)}{\partial x}. \qquad (14.28)$$

This equation is analogous to the classical equation, the Newton's second law

$$\frac{dp_x}{dt} = F_x = -\frac{\partial V(x)}{\partial x}, \qquad (14.29)$$

where F_x is the force acting on the particle.

14.6 Interaction Picture

We have already demonstrated that when the Hamiltonian of a given system is independent of time, either the states or operators can depend explicitly on time. We have distinguished those two cases, respectively, as the Schrödinger and Heisenberg pictures. We have also seen that the time dependence can be transferred, using a unitary transformation, from the vectors to the operators, and vice versa.

When, however, the Hamiltonian $\hat{H}_t$ depends explicitly on time, the situation is more complicated that both the state vectors and operators can, in general, depend simultaneously on time and there is no unitary transformation that could transfer them to either the Schrödinger or Heisenberg pictures.

In many situations in physics, it is possible to write the time-dependent Hamiltonian $\hat{H}_t$ as a sum of two terms

$$\hat{H}_t = \hat{H}_0 + \hat{V}(t), \qquad (14.30)$$

in which only the second term $\hat{V}(t)$ depends explicitly on time. In particular, this can happen in composite systems composed of two or more subsystems. Usually, in this case the time-dependent part of the Hamiltonian represents the interaction energy between subsystems.

In this case, we can define a unitary operator involving only the time-independent part of the Hamiltonian

$$\hat{U}_0(t, t_0) = e^{-\frac{i}{\hbar}\hat{H}_0(t-t_0)}, \qquad (14.31)$$

and a transformed state

$$|\Psi_I(t)\rangle \equiv |\Psi(t)\rangle_T = \hat{U}_0^\dagger(t, t_0)|\Psi_s(t)\rangle. \tag{14.32}$$

Using this transformation, we can transform the time-dependent Schrödinger equation for $|\Psi_s(t)\rangle$, given in the Schrödinger picture, into a time-dependent equation for the state vector $|\Psi_I(t)\rangle$, which we shall call the Schrödinger equation in the *interaction picture*. Beginning from Eq. (14.14), which is the Schrödinger equation for the transformed state and using Eq. (14.30), we get

$$i\hbar\frac{\partial}{\partial t}|\Psi_I(t)\rangle = \left[\hat{U}_0^\dagger\left(\hat{H}_0 + \hat{V}(t)\right)\hat{U}_0 - \hat{H}_0\right]|\Psi_I(t)\rangle. \tag{14.33}$$

Since $\hat{U}_0^\dagger \hat{H}_0 \hat{U}_0 = \hat{H}_0$, we then have

$$i\hbar\frac{\partial}{\partial t}|\Psi_I(t)\rangle = \hat{U}_0^\dagger \hat{V}(t)\hat{U}_0|\Psi_I(t)\rangle = \hat{V}_I(t)|\Psi_I(t)\rangle, \tag{14.34}$$

where $\hat{V}_I(t) = \hat{U}_0^\dagger \hat{V}(t)\hat{U}_0$. Hence, the evolution of the state vector $|\Psi_I(t)\rangle$ is solely determined by the time-dependent part of the Hamiltonian, the interaction part $\hat{V}(t)$. For this reason, the vector $|\Psi_I(t)\rangle$ is called the state vector in the interaction picture.

When, for example, $\hat{V}_I$ does not depend explicitly on time, we can write

$$|\Psi_I(t)\rangle = \hat{U}_I(t, t_0)|\Psi_I(t_0)\rangle, \tag{14.35}$$

where

$$\hat{U}_I(t, t_0) = e^{-\frac{i}{\hbar}\hat{V}_I(t-t_0)} \tag{14.36}$$

is the evolution operator in the interaction picture. Since

$$|\Psi_I(t_0)\rangle = |\Psi_s(t_0)\rangle = \hat{U}^\dagger(t, t_0)|\Psi_s(t)\rangle, \tag{14.37}$$

we get

$$|\Psi_I(t)\rangle = \hat{U}_I(t, t_0)\hat{U}^\dagger(t, t_0)|\Psi_s(t)\rangle. \tag{14.38}$$

Comparing Eq. (14.38) with Eq. (14.32), we see that

$$\hat{U}_0^\dagger(t, t_0) = \hat{U}_I(t, t_0)\hat{U}^\dagger(t, t_0), \tag{14.39}$$

or

$$\hat{U}_I(t, t_0) = \hat{U}_0^\dagger(t, t_0)\hat{U}(t, t_0). \tag{14.40}$$

Substituting Eq. (14.35) into Eq. (14.34), we get the equation of motion for the unitary operator $\hat{U}_I(t, t_0)$:

$$i\hbar\frac{d\hat{U}_I}{dt} = \hat{V}_I\hat{U}_I. \qquad (14.41)$$

Similar to the state vector $|\Psi_I(t)\rangle$, the evolution of the unitary operator in the interaction picture is determined by the interaction Hamiltonian $\hat{V}$.

In summary, using the unitary operator $\hat{U}_0(t, t_0)$ involving only the time-independent part of the Hamiltonian, we can transform the state vector of a given system into the interaction picture in which the evolution if the transformed state is determined solely by the time-dependent (interaction) part of the Hamiltonian.

What about the evolution of an operator $\hat{A}(t)$? Define a transformed operator

$$\hat{A}_I(t) = \hat{U}_I^\dagger\hat{A}(t)\hat{U}_I, \qquad (14.42)$$

and find its time evolution

$$\frac{d}{dt}\hat{A}_I(t) = \left(\frac{\partial}{\partial t}\hat{U}_I^\dagger\right)\hat{A}(t)\hat{U}_I + \hat{U}_I^\dagger\left(\frac{\partial}{\partial t}\hat{A}(t)\right)\hat{U}_I + \hat{U}_I^\dagger\hat{A}(t)\left(\frac{\partial}{\partial t}\hat{U}_I\right)$$

$$= \frac{i}{\hbar}\hat{U}_I^\dagger\hat{V}_I\hat{A}(t)\hat{U}_I + \hat{U}_I^\dagger\left(\frac{\partial}{\partial t}\hat{A}(t)\right)\hat{U}_I - \frac{i}{\hbar}\hat{U}_I^\dagger\hat{A}(t)\hat{V}_I\hat{U}_I$$

$$= \frac{i}{\hbar}\hat{U}_I^\dagger\hat{V}_I\hat{U}_I\hat{U}_I^\dagger\hat{A}(t)\hat{U}_I + \hat{U}_I^\dagger\left(\frac{\partial}{\partial t}\hat{A}(t)\right)\hat{U}_I$$

$$\quad - \frac{i}{\hbar}\hat{U}_I^\dagger\hat{A}(t)\hat{U}_I\hat{U}_I^\dagger\hat{V}_I\hat{U}_I$$

$$= \frac{i}{\hbar}\hat{V}_I\hat{A}_I + \hat{U}_I^\dagger\left(\frac{\partial}{\partial t}\hat{A}(t)\right)\hat{U}_I - \frac{i}{\hbar}\hat{A}_I\hat{V}_I$$

$$= \hat{U}_I^\dagger\left(\frac{\partial}{\partial t}\hat{A}(t)\right)\hat{U}_I + \frac{i}{\hbar}[\hat{V}_I, \hat{A}_I]. \qquad (14.43)$$

Hence, the operator $\hat{A}_I(t)$ evolves in time even if the operator $\hat{A}(t)$ does not explicitly depend on time. The evolution of the operator is solely determined by the interaction Hamiltonian $\hat{V}_I$. For this reason, the operator $\hat{A}_I(t)$ is called the operator of a given system in the interaction picture.

14.7 Ehrenfest Theorem

The Ehrenfest theorem says that quantum mechanics must produce the same result as the classical mechanics for a system in which the particle can be represented by a well-localized wave function, i.e., the wave function with $\partial \Psi(x, t)/\partial x = 0$. Thus, the Ehrenfest theorem is an example of the correspondence principle.

According to the Ehrenfest theorem, the equations of motion for the average (mean or expectation) values of the position and momentum operators of a particle, which can be represented by well-localized wave functions, are identical to the classical equations of motion. Consider the equations of motion for the average values of the position $\hat{x}$ and momentum $\hat{p}_x$ operators of a particle of mass m moving in a one-dimensional potential $\hat{V}$:

$$\frac{d}{dt}\langle \hat{x}\rangle = \frac{d}{dt}\langle \Psi_H|\hat{x}|\Psi_H\rangle = \langle \Psi_H|\frac{d}{dt}\hat{x}|\Psi_H\rangle$$

$$= \frac{1}{m}\langle \Psi_H|\hat{p}_x|\Psi_H\rangle = \frac{1}{m}\langle \hat{p}_x\rangle, \tag{14.44}$$

$$\frac{d}{dt}\langle \hat{p}_x\rangle = \frac{d}{dt}\langle \Psi_H|\hat{p}_x|\Psi_H\rangle = \langle \Psi_H|\frac{d}{dt}\hat{p}_x|\Psi_H\rangle = -\left\langle \frac{\partial \hat{V}}{\partial x}\right\rangle. \tag{14.45}$$

Equation (14.45) can be written as

$$m\frac{d^2}{dt^2}\langle \hat{x}\rangle = -\left\langle \frac{\partial \hat{V}}{\partial x}\right\rangle. \tag{14.46}$$

However, classical physics says that the right-hand side of Eq. (14.46) is a force F_x not $\langle F_x\rangle$. To solve the problem, we can use the assumption that the particle is represented by the well-localized wave function. Then

$$\left\langle \frac{\partial \hat{V}}{\partial x}\right\rangle = \int \Psi^*(x, t)\frac{\partial \hat{V}}{\partial x}\Psi(x, t)dx. \tag{14.47}$$

Note that

$$\frac{\partial \hat{V}}{\partial x}\Psi(x, t) = \left(\frac{\partial \hat{V}}{\partial x}\right)\Psi(x, t) + \hat{V}\frac{\partial \Psi(x, t)}{\partial x}. \tag{14.48}$$

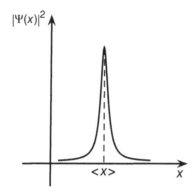

Figure 14.1 Example of a well-localized wave function. The function is different from zero only in a very narrow region around the expectation value of x.

Thus, if $\Psi(x, t)$ is a well-localized function, then $\partial \Psi(x, t)/\partial x = 0$, and we can write Eq. (14.47) as

$$\left\langle \frac{\partial \hat{V}}{\partial x} \right\rangle = \int \Psi^*(x, t) \frac{\partial \hat{V}}{\partial x} \Psi(x, t) dx = \int \Psi^*(x, t) \left(\frac{\partial \hat{V}}{\partial x} \right) \Psi(x, t) dx$$

$$\approx \left(\frac{\partial \hat{V}}{\partial x} \right)_{\langle x \rangle} \int |\Psi(x, t)|^2 dx = \left(\frac{\partial \hat{V}}{\partial x} \right)_{\langle x \rangle}, \qquad (14.49)$$

where $(\partial \hat{V}/\partial x)_{\langle x \rangle}$ is the value of the derivative at the maximum of the wave function.

Figure 14.1 shows an example of a well-localized wave function that is maximal at the expectation value of x and is different from zero only in a very narrow region around $\langle x \rangle$. Thus, a measurement of x is almost certain to yield a result, which is very close to $\langle x \rangle$. The best shape for the wave function for a perfect correspondence between the quantum and classical mechanics would be a δ function, $\Psi(x) = \delta(x - \langle x \rangle)$.

In summary, quantum mechanics produces the same result as classical mechanics for a system in which particles are represented by well-localized (narrow) wave functions. In this case, the expectation values of the operators correspond to the classical limits of the observables, which is consistent with the correspondence principle.

Tutorial Problems

Problem 14.1 Consider a two-level atom of energy states $|1\rangle$ and $|2\rangle$ driven by a laser field. The atom can be represented as a spin-$\frac{1}{2}$ particle and the laser field can be treated as a classical field. The Hamiltonian of the system is given by

$$\hat{H} = \frac{1}{2}\hbar\omega_0\hat{\sigma}_z - \frac{1}{2}i\hbar\Omega\left(\hat{\sigma}^+ e^{-i\omega_L t} - \hat{\sigma}^- e^{i\omega_L t}\right), \qquad (14.50)$$

where Ω is the Rabi frequency of the laser field, ω_0 is the atomic transition frequency, ω_L is the laser frequency, and $\hat{\sigma}_z$, $\hat{\sigma}^+$, and $\hat{\sigma}^-$ are the spin operators defined as

$$\hat{\sigma}_z = |2\rangle\langle 2| - |1\rangle\langle 1|, \quad \hat{\sigma}^+ = |2\rangle\langle 1|, \quad \hat{\sigma}^- = |1\rangle\langle 2|. \quad (14.51)$$

(a) Calculate the equation of motion for $\hat{\sigma}^-$.

(b) The equation of motion derived in (a) contains a time-dependent coefficient. Find a unitary operator that transforms $\hat{\sigma}^-$ into $\hat{\tilde{\sigma}}^-$ whose equation of motion is free of time-dependent coefficients.

Problem 14.2 The Hamiltonian of the two-level atom interacting with a classical laser field can be written as

$$\hat{H} = \hat{H}_0 + \hat{V}(t), \qquad (14.52)$$

where

$$\hat{H}_0 = \frac{1}{2}\hbar\omega_0\hat{\sigma}_z$$

$$\hat{V}(t) = -\frac{1}{2}i\hbar\Omega\left(\hat{\sigma}^+ e^{-i\omega_L t} - \hat{\sigma}^- e^{i\omega_L t}\right). \qquad (14.53)$$

(a) Transform $\hat{V}(t)$ into the interaction picture to find $\hat{V}_I = \hat{U}_0^\dagger \hat{V}(t)\hat{U}_0$.

(b) Find the equation of motion for $\hat{\sigma}^-$ in the interaction picture, i.e., find the equation of motion for $\hat{\sigma}_I^-(t) = \hat{U}_I^\dagger \hat{\sigma}^- \hat{U}_I$.

Chapter 15

Quantum Harmonic Oscillator

We have illustrated in Section 8.3 the solution to the stationary Schrödinger equation for a particle in a square-well potential, where $V(x)$ had a special simple structure (step function).

In this chapter, we shall try to investigate a more complicated system: the one-dimensional Schrödinger equation with the potential $V(x)$ strongly dependent on x, such that

$$\hat{V}(x) = \frac{1}{2}m\omega^2\hat{x}^2 . \tag{15.1}$$

Readers familiar with harmonic motion will recognize that this is the well-known potential of a one-dimensional harmonic oscillator of mass m and frequency of oscillations ω. Its dependence on the amplitude of the oscillation x is illustrated in Fig. 15.1. This is still a one-dimensional problem but with a complication arising from the x dependence of the potential.

The study of quantum properties of the harmonic oscillator is important in physics as many real-world systems oscillate harmonically. Motion of systems in a confined space is often modeled as being a quantized harmonic motion or, in first instance, is approximated by a harmonic motion.

Quantum Physics for Beginners
Zbigniew Ficek
Copyright © 2016 Pan Stanford Publishing Pte. Ltd.
ISBN 978-981-4669-38-2 (Hardcover), 978-981-4669-39-9 (eBook)
www.panstanford.com

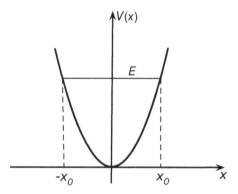

Figure 15.1 Potential of a one-dimensional harmonic oscillator. The amplitude x_0 of the motion is called the classical turning point and is determined by the total energy E of the oscillator, which classically can have any value.

Consider a harmonic oscillator composed of an oscillating mass m. In one dimension, the Hamiltonian of the harmonic oscillator is given by

$$\hat{H} = \frac{1}{2m}\hat{p}^2 + \frac{1}{2}m\omega^2\hat{x}^2 , \qquad (15.2)$$

where $\hat{p}^2/2m$ is the kinetic energy and $m\omega^2\hat{x}^2/2$ is the potential energy of the mass.

We will find energies (eigenvalues) and eigenfunctions of the harmonic oscillator by solving the stationary Schrödinger equation (eigenvalue equation) using two different approaches.

In the first, we will solve the equation employing an algebraic operator technique, which is based on the Dirac notation. This approach has several definite advantages and exploits the commutation relations among the operators involved and their properties.

In the second approach, we will transform the stationary Schrödinger equation into a second-order differential equation and will find the solution to the equation with the aid of a more advanced mathematical technique that involves *special functions*.

15.1 Algebraic Operator Technique

The algebraic operator technique is based on the commutation relation of two Hermitian operators involved in the evolution of the harmonic oscillator: position $\hat{x}$ and momentum $\hat{p} = \hat{p}_x$:

$$[\hat{x}, \hat{p}] = i\hbar. \tag{15.3}$$

We will introduce a non-Hermitian operator defined as

$$\hat{a} = \sqrt{\frac{m\omega}{2\hbar}}\hat{x} + i\frac{1}{\sqrt{2m\hbar\omega}}\hat{p}, \tag{15.4}$$

and the adjoint of this operator

$$\hat{a}^\dagger = \sqrt{\frac{m\omega}{2\hbar}}\hat{x} - i\frac{1}{\sqrt{2m\hbar\omega}}\hat{p}. \tag{15.5}$$

Using the commutation relation (15.3), we find that the operators $\hat{a}$, $\hat{a}^\dagger$ satisfy the commutation relation

$$[\hat{a}, \hat{a}^\dagger] = \hat{1}. \tag{15.6}$$

This allows us to write the Hamiltonian $\hat{H}$ in a compact form

$$\hat{H} = \frac{1}{2}\hbar\omega\left(\hat{a}^\dagger\hat{a} + \hat{a}\hat{a}^\dagger\right) = \hbar\omega\left(\hat{a}^\dagger\hat{a} + \frac{1}{2}\right). \tag{15.7}$$

Hence, the eigenvalue equation

$$\hat{H}|\phi\rangle = E|\phi\rangle, \tag{15.8}$$

can be written as

$$\hbar\omega\left(\hat{a}^\dagger\hat{a} + \frac{1}{2}\right)|\phi\rangle = E|\phi\rangle. \tag{15.9}$$

Multiplying Eq. (15.9) from the left by $\langle\phi|$, and using the normalization $\langle\phi|\phi\rangle = 1$, we get

$$\hbar\omega\left(\langle\phi|\hat{a}^\dagger\hat{a}|\phi\rangle + \frac{1}{2}\right) = E. \tag{15.10}$$

Since

$$\langle\phi|\hat{a}^\dagger\hat{a}|\phi\rangle = (\hat{a}|\phi\rangle, \hat{a}|\phi\rangle) \geq 0, \tag{15.11}$$

we have that

$$E \geq \frac{1}{2}\hbar\omega. \tag{15.12}$$

Thus, the energy of the quantum harmonic oscillator can never be zero.

From Eq. (15.9), we can generate a new eigenvalue equation multiplying this equation from the left by $\hat{a}^\dagger$:

$$\hbar\omega \left(\hat{a}^\dagger \hat{a}^\dagger \hat{a} + \frac{1}{2}\hat{a}^\dagger \right) |\phi\rangle = E \hat{a}^\dagger |\phi\rangle . \tag{15.13}$$

Using the commutation relation (15.6), we can write Eq. (15.13) as

$$\hbar\omega \left(\hat{a}^\dagger \hat{a} - \frac{1}{2} \right) \hat{a}^\dagger |\phi\rangle = E \hat{a}^\dagger |\phi\rangle . \tag{15.14}$$

Adding to both sides $\hbar\omega \hat{a}^\dagger |\phi\rangle$, we obtain

$$\hbar\omega \left(\hat{a}^\dagger \hat{a} + \frac{1}{2} \right) \hat{a}^\dagger |\phi\rangle = (E + \hbar\omega) \hat{a}^\dagger |\phi\rangle . \tag{15.15}$$

Introducing a notation $|\Psi\rangle = \hat{a}^\dagger |\phi\rangle$, we see that $|\Psi\rangle$ is an eigenfunction of $\hat{H}$ with eigenvalue $E + \hbar\omega$.

Thus, the operator $\hat{a}^\dagger$ acting on the state $|\phi\rangle$ of energy E transforms this state to the state $|\Psi\rangle$ of energy $E + \hbar\omega$. Therefore, the operator $\hat{a}^\dagger$ is called the *raising operator* or *creation operator*.

Now, multiplying Eq. (15.15) from the left by $\hat{a}^\dagger$, we obtain

$$\hbar\omega \left(\hat{a}^\dagger \hat{a}^\dagger \hat{a} + \frac{1}{2}\hat{a}^\dagger \right) |\Psi\rangle = (E + \hbar\omega) \hat{a}^\dagger |\Psi\rangle . \tag{15.16}$$

Proceeding similar as above, we get

$$\hbar\omega \left(\hat{a}^\dagger \hat{a} + \frac{1}{2} \right) \hat{a}^\dagger |\Psi\rangle = (E + 2\hbar\omega) \hat{a}^\dagger |\Psi\rangle . \tag{15.17}$$

Thus, the state $\hat{a}^\dagger |\Psi\rangle = \hat{a}^\dagger \hat{a}^\dagger |\phi\rangle$ is an eigenfunction of $\hat{H}$ with an eigenvalue $E + 2\hbar\omega$.

Similarly, we can show that the state $|\phi_n\rangle = (\hat{a}^\dagger)^n |\phi\rangle$ is an eigenfunction of $\hat{H}$ with an eigenvalue $E + n\hbar\omega$.

Consider now the action of the operator $\hat{a}$ on the eigenfunctions to find the resulting eigenvalues and eigenfunctions.

Take the eigenvalue equation for $|\phi_n\rangle$:

$$\hbar\omega \left(\hat{a}^\dagger \hat{a} + \frac{1}{2} \right) |\phi_n\rangle = (E + n\hbar\omega) |\phi_n\rangle = E_n |\phi_n\rangle . \tag{15.18}$$

Multiplying Eq. (15.18) from the left by $\hat{a}$, we get

$$\hbar\omega \left(\hat{a} \hat{a}^\dagger \hat{a} + \frac{1}{2}\hat{a} \right) |\phi_n\rangle = (E + n\hbar\omega) \hat{a} |\phi_n\rangle , \tag{15.19}$$

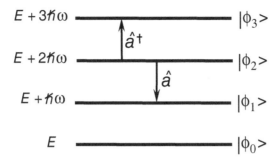

Figure 15.2 Energy spectrum of the harmonic oscillator and the action of the annihilation and creation operators. Note that the discrete energy levels are equally separated from each other by $\hbar\omega$.

and using the commutation relation (15.6), we obtain

$$\hbar\omega \left(\hat{a}^\dagger \hat{a} \hat{a} + \frac{3}{2} \hat{a} \right) |\phi_n\rangle = (E + n\hbar\omega) \, \hat{a} |\phi_n\rangle . \qquad (15.20)$$

Hence

$$\hbar\omega \left(\hat{a}^\dagger \hat{a} + \frac{1}{2} \right) \hat{a} |\phi_n\rangle = [E + (n-1)\hbar\omega] \, \hat{a} |\phi_n\rangle . \qquad (15.21)$$

Thus, the state $|\phi_{n-1}\rangle = \hat{a} |\phi_n\rangle$ is an eigenfunction of $\hat{H}$ with an eigenvalue $E_n - \hbar\omega$. Therefore, the operator $\hat{a}$ is called the *lowering operator* or *annihilation operator*.

Suppose that the state $|\phi_0\rangle$ of energy E is the lowest (ground) state of the harmonic oscillator. Thus, the energy spectrum (eigenvalues), shown in Fig. 15.2, forms a ladder of equally spaced levels separated by $\hbar\omega$, which one ascends by the action of $\hat{a}^\dagger$ and descends by the action of $\hat{a}$. The quantum harmonic oscillator, therefore, has a **discrete** energy spectrum.

Consider the action of $\hat{a}$ on the ground state

$$\hbar\omega \left(\hat{a}^\dagger \hat{a} + \frac{1}{2} \right) \hat{a} |\phi_0\rangle = (E - \hbar\omega) \, \hat{a} |\phi_0\rangle . \qquad (15.22)$$

This equation cannot be satisfied. Otherwise, there would exist another eigenvalue $E - \hbar\omega$ lower than E. Thus, $\hat{a} |\phi_0\rangle$ must be identically zero:

$$\hat{a} |\phi_0\rangle \equiv 0 . \qquad (15.23)$$

Hence, the eigenvalue equation for the ground state is

$$\hat{H}|\phi_0\rangle = \hbar\omega\left(\hat{a}^\dagger\hat{a} + \frac{1}{2}\right)|\phi_0\rangle = \frac{1}{2}\hbar\omega|\phi_0\rangle. \qquad (15.24)$$

Thus, the energy (eigenvalue) of the ground state is $E = \hbar\omega/2$.

We can summarize our findings that the energy eigenvalues of the harmonic oscillator are discrete

$$E_n = \left(n + \frac{1}{2}\right)\hbar\omega, \qquad n = 0, 1, 2, \ldots \qquad (15.25)$$

with corresponding eigenfunctions

$$|\phi_0\rangle, \quad |\phi_1\rangle = \hat{a}^\dagger|\phi_0\rangle, \quad |\phi_2\rangle = \left(\hat{a}^\dagger\right)^2|\phi_0\rangle, \ldots, |\phi_n\rangle = \left(\hat{a}^\dagger\right)^n|\phi_0\rangle. \qquad (15.26)$$

From the above equation, it follows that starting with $|\phi_0\rangle$, we may obtain the complete set of eigenvectors of the harmonic oscillator by repeatedly applying the operator $\hat{a}^\dagger$ on the eigenstate $|\phi_0\rangle$.

However, the eigenstates found in this way are not normalized. The normalization of $\phi_n(x) = c_n\left(\hat{a}^\dagger\right)^n\phi_0(x)$ gives

$$\begin{aligned}
1 = \langle\phi_n|\phi_n\rangle &= |c_n|^2\langle\phi_0|\left(\hat{a}^{\dagger n}\right)^\dagger\left(\hat{a}^\dagger\right)^n|\phi_0\rangle \\
&= |c_n|^2\langle\phi_0|\hat{a}^n\hat{a}^{\dagger n}|\phi_0\rangle \\
&= |c_n|^2\langle\phi_0|\hat{a}^{n-1}\hat{a}\hat{a}^{\dagger n}|\phi_0\rangle. \qquad (15.27)
\end{aligned}$$

Using the commutation relation[a]

$$\left[\hat{a}, \left(\hat{a}^\dagger\right)^n\right] = n\left(\hat{a}^\dagger\right)^{n-1}, \qquad (15.28)$$

we can continue Eq. (15.27) as

$$\begin{aligned}
&= |c_n|^2\langle\phi_0|\hat{a}^{n-1}\left(n\hat{a}^{\dagger n-1} + \hat{a}^{\dagger n}\hat{a}\right)|\phi_0\rangle \\
&= |c_n|^2 n\langle\phi_0|\hat{a}^{n-1}\hat{a}^{\dagger n-1}|\phi_0\rangle \\
&= |c_n|^2 n\langle\phi_0|\hat{a}^{n-2}\left((n-1)\hat{a}^{\dagger n-2} + \hat{a}^{\dagger n-1}\hat{a}\right)|\phi_0\rangle \\
&= |c_n|^2 n(n-1)\langle\phi_0|\hat{a}^{n-2}\hat{a}^{\dagger n-2}|\phi_0\rangle. \qquad (15.29)
\end{aligned}$$

Proceeding further, we find that Eq. (15.27) reduces to

$$1 = |c_n|^2 n!. \qquad (15.30)$$

---

[a]Proof by induction left for the readers as a tutorial problem.

Thus, the normalized eigenfunctions of the harmonic oscillator are

$$|\phi_n\rangle = \frac{1}{\sqrt{n!}} \left(\hat{a}^{\dagger}\right)^n |\phi_0\rangle . \tag{15.31}$$

Equation (15.31) shows that an nth eigenfunction can be generated from the ground state eigenfunction by the n-times repeated action of the creation operator on $|\phi_0\rangle$. Thus, it is enough to know the ground state eigenfunction to find all the eigenfunctions of the harmonic oscillator.

This is the complete solution to the problem. It is remarkable that the commutation relation (15.6) was all what we needed to deal with the harmonic oscillator completely. In a very effective way, we extracted the essential structure of the problem and have founded the eigenvalues and eigenvectors of the harmonic oscillator.

Using the definition of the ground state (15.23), we may find the explicit form of the ground state eigenfunction. Substituting for $\hat{a}$ from Eq. (15.4) and using the explicit form of $\hat{p} = -i\hbar d/dx$, we get

$$\sqrt{\frac{m\omega}{2\hbar}} x\phi_0 + i \frac{1}{\sqrt{2\hbar m\omega}} \left(-i\hbar \frac{d\phi_0}{dx}\right) = 0 , \tag{15.32}$$

which simplifies to

$$\frac{d\phi_0}{dx} + \frac{m\omega}{\hbar} x\phi_0 = 0 , \tag{15.33}$$

where $\phi_0 \equiv |\phi_0\rangle$. Hence

$$\frac{d\phi_0}{\phi_0} = -\frac{m\omega}{\hbar} x dx . \tag{15.34}$$

Integrating Eq. (15.34), we obtain

$$\ln \frac{\phi_0(x)}{\phi_0(0)} = -\frac{m\omega}{2\hbar} x^2 , \tag{15.35}$$

from which we find

$$\phi_0(x) = \phi_0(0) \exp\left(-\frac{m\omega}{2\hbar} x^2\right) . \tag{15.36}$$

We find $\phi_0(0)$ from the normalization, which finally gives

$$\phi_0(x) = \left(\frac{m\omega}{\pi\hbar}\right)^{\frac{1}{4}} \exp\left(-\frac{m\omega}{2\hbar} x^2\right) . \tag{15.37}$$

Thus, the wave function of the ground state is a Gaussian.

The wave functions $\phi_n(x)$ of the other states can be found from the relation

$$\phi_n(x) = \left(\hat{a}^\dagger\right)^n \phi_0(x).$$ (15.38)

Using the definition of $\hat{a}^\dagger$ (Eq. (15.5)), we can find $\phi_n(x)$ in terms of the position x:

$$\phi_1(x) = \hat{a}^\dagger \phi_0(x)$$
$$= \left[\sqrt{\frac{m\omega}{2\hbar}}x - i\frac{1}{\sqrt{2m\hbar\omega}}\left(-i\hbar\frac{d}{dx}\right)\right]\phi_0(x).$$ (15.39)

From Eq. (15.33), we have

$$\frac{d\phi_0}{dx} = -\frac{m\omega}{\hbar}x\phi_0.$$ (15.40)

Hence

$$\phi_1(x) = \left(\sqrt{2}\sqrt{\frac{m\omega}{\hbar}}x\right)\phi_0(x).$$ (15.41)

Similarly, we can find that

$$\phi_2(x) = \frac{1}{\sqrt{2}}\left[2\left(\frac{m\omega}{\hbar}\right)x^2 - 1\right]\phi_0(x).$$ (15.42)

We can introduce a new parameter

$$\alpha = \sqrt{\frac{m\omega}{\hbar}}x,$$ (15.43)

and write the wave functions as

$$\phi_1(\alpha) = \frac{1}{\sqrt{2}}H_1(\alpha)\phi_0(\alpha), \quad \phi_2(\alpha) = \frac{1}{2\sqrt{2}}H_2(\alpha)\phi_0(\alpha),$$ (15.44)

where $H_n(\alpha)$ are Hermite polynomials of degree n.
First few Hermite polynomials

$$H_0(\alpha) = 1, \quad H_1(\alpha) = 2\alpha, \quad H_2(\alpha) = 4\alpha^2 - 2, \ldots$$ (15.45)

Hermite polynomials satisfy the differential equation

$$\frac{d^2 H_n(\alpha)}{d\alpha^2} - 2\alpha\frac{d H_n(\alpha)}{d\alpha} + 2n H_n(\alpha) = 0.$$ (15.46)

Notice that the wave functions of the harmonic oscillator are not in the form of sinusoidal functions.

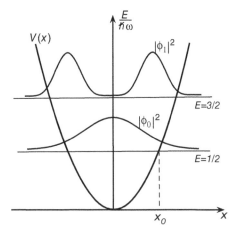

Figure 15.3 First two energy eigenvalues and the corresponding eigenfunctions of the harmonic oscillator.

Let us consider in more details the properties of the harmonic oscillator being in its ground state. Using the classical representation of energy, we have

$$\frac{1}{2}\hbar\omega = \frac{p^2}{2m} + \frac{1}{2}m\omega^2 x^2 . \tag{15.47}$$

Since, $p^2 \geq 0$, the particle must be restricted to positions x, such that

$$\frac{1}{2}m\omega^2 x^2 \leq \frac{1}{2}\hbar\omega , \tag{15.48}$$

i.e.,

$$|x| \leq \sqrt{\frac{\hbar}{m\omega}} . \tag{15.49}$$

The maximum of $|x| \equiv x_0 = \sqrt{\hbar/m\omega}$ is called the *classical turning point*.

Since the wave function $\psi_0(x)$ is not restricted to $x \leq x_0$, see Fig. 15.3, quantum mechanics predicts that the harmonic oscillator can be in the classically forbidden region.

Another interesting observation: According to classical theory of harmonic oscillator, the probability of finding the oscillating mass at a given position is greatest at the end points of its motion, where it moves slowly, and least near the equilibrium position ($x = 0$), where it moves rapidly.

Exactly opposite behavior is manifested by a quantum oscillator in its lowest energy state of $n = 0$. The probability density has its maximum value at $x = 0$ and drops off on either side of the position.

15.2 Special Functions Method

We will carry out the solution to the eigenvalue equation of the harmonic oscillator again, this time using the stationary Schrödinger equation in a form of a second-order differential equation.

The starting point is the stationary Schrödinger equation for the harmonic oscillator whose Hamiltonian is of the form

$$\hat{H} = \frac{1}{2m}\hat{p}^2 + \frac{1}{2}m\omega^2\hat{x}^2 . \tag{15.50}$$

Since in one dimension, $\hat{p} = -i\hbar d/dx$, the Schrödinger (eigenvalue) equation takes the form

$$\left(-\frac{\hbar^2}{2m}\frac{d^2}{dx^2} + \frac{1}{2}m\omega^2 x^2\right)\phi = E\phi , \tag{15.51}$$

or multiplying by $-2m$ and dividing by $\hbar^2$, we obtain a second-order differential equation

$$\frac{d^2\phi}{dx^2} + \frac{2m}{\hbar^2}\left(E - \frac{m\omega^2}{2}x^2\right)\phi = 0 . \tag{15.52}$$

This is not a linear differential equation, and it is not easy to obtain a solution.

We can proceed in the following way. Introducing new variables

$$\lambda = \frac{2m}{\hbar^2}E , \qquad \beta^2 = \frac{m^2\omega^2}{\hbar^2} , \tag{15.53}$$

we can write Eq. (15.52) in a simpler form

$$\frac{d^2\phi}{dx^2} + \left(\lambda - \beta^2 x^2\right)\phi = 0 , \tag{15.54}$$

which is still difficult to solve. Despite the difficulty, we will try to solve the differential equation (15.54). First, we will find the solution to Eq. (15.54) in the asymptotic limit of large x ($x \gg 1$). In this limit, we can ignore the λ term as being small compared to $\beta^2 x^2$ and obtain

$$\frac{d^2\phi}{dx^2} - \beta^2 x^2\phi = 0 . \tag{15.55}$$

This is a simple differential equation whose solution is of the form

$$\phi(x) = C \exp\left(-\frac{1}{2}\beta x^2\right), \qquad (15.56)$$

where C is a constant.

Hence, we will try to find the solution to Eq. (15.54) in the form

$$\phi(x) = f(x) \exp\left(-\frac{1}{2}\beta x^2\right), \qquad (15.57)$$

i.e., in the form satisfying the asymptotic solution (15.56), where $f(x)$ is a function of x, which remains to be found.

By substituting Eq. (15.57) into Eq. (15.54), we obtain

$$\frac{d^2 f}{dx^2} - 2\beta x \frac{df}{dx} + (\lambda - \beta) f = 0. \qquad (15.58)$$

Introducing a new variable $\alpha = \sqrt{\beta} x$ and a new function $f(x) \to H(\alpha)$, for which

$$\frac{df}{dx} = \frac{dH}{d\alpha}\frac{d\alpha}{dx} = \sqrt{\beta}\frac{dH}{d\alpha},$$

$$\frac{d^2 f}{dx^2} = \sqrt{\beta}\frac{d^2 H}{d\alpha^2}\frac{d\alpha}{dx} = \beta\frac{d^2 H}{d\alpha^2}, \qquad (15.59)$$

the differential equation (15.58) transforms into

$$\frac{d^2 H}{d\alpha^2} - 2\alpha\frac{dH}{d\alpha} + \left(\frac{\lambda}{\beta} - 1\right) H = 0. \qquad (15.60)$$

Having written the Schrödinger equation in this form, one can notice that it is identical to the differential equation for Hermite polynomials, Eq. (15.46), with

$$\frac{\lambda}{\beta} - 1 = 2n, \qquad (15.61)$$

where n is integer.[a]

---

[a] Of course, we can prove by using the standard procedure of solving differential equations that the solution to Eq. (15.60) is really in the form of Hermite polynomials. We leave the details of the solution as a tutorial problem.

Thus, the wave functions of the harmonic oscillator are of the form

$$\phi_n(x) = N H_n(\alpha) \exp\left(-\frac{1}{2}\alpha^2\right),$$
(15.62)

where N is a normalization constant.

Since n is integer, we find from Eqs. (15.61) and (15.53) that the energy eigenvalue E is

$$E = \left(n + \frac{1}{2}\right)\hbar\omega.$$
(15.63)

In summary, from the foregoing treatment of the harmonic oscillator, we see that the solution to the Schrödinger equation given in the differential form agrees perfectly with the results obtained by the algebraic operator technique.

In summary of this chapter, we have learned that

(1) The energy of a harmonic oscillator is quantized, with the sequence of values

$$E_n = \left(n + \frac{1}{2}\right)\hbar\omega, \qquad n = 0, 1, 2, \ldots$$

(2) The energy levels are equally spaced. This is an important point to remember. The difference in energy between adjacent energy levels is equal to the energy of a single photon, $\hbar\omega$.
(3) The lowest energy the oscillator can have is $E_0 = \frac{1}{2}\hbar\omega$, which is nonzero. Thus, the oscillator can never be made stationary.
(4) The oscillator can be found in the classically forbidden region. This is another example of penetration of a potential barrier or quantum tunneling.

Worked Example

Assume that the harmonic oscillator is in the ground state $n = 0$. Calculate the probability that the oscillator will be found in the classically forbidden region, where the kinetic energy is negative.

Solution

We have shown in the chapter that the wave function of the ground state is

$$\phi_0(x) = Ae^{-\beta x^2},$$

where

$$A = \left(\frac{m\omega}{\pi\hbar}\right)^{\frac{1}{4}} \quad \text{and} \quad \beta = \frac{m\omega}{2\hbar}.$$

Classically forbidden regions are $x \leq -x_0$ and $x \geq x_0$, where $x_0 = \sqrt{\hbar/m\omega}$ is the classical turning point (see Fig. 15.3).

Probability of finding the harmonic oscillator in the classically forbidden region is

$$P = \int_{-\infty}^{-x_0} |\phi_0(x)|^2 dx + \int_{x_0}^{\infty} |\phi_0(x)|^2 dx$$

$$= 2A^2 \int_{x_0}^{\infty} e^{-2\beta x^2} dx = 2A^2 \int_{\frac{1}{\sqrt{2\beta}}}^{\infty} e^{-2\beta x^2} dx.$$

By substituting

$$y^2 = 2\beta x^2,$$

we can change the variable

$$x = \frac{1}{\sqrt{2\beta}} y \quad \text{and} \quad dx = \frac{1}{\sqrt{2\beta}} dy.$$

Hence, we find

$$P = \frac{2A^2}{\sqrt{2\beta}} \int_1^{\infty} e^{-y^2} dy = \frac{2}{\sqrt{\pi}} \int_1^{\infty} e^{-y^2} dy = 1 - \text{Erf}(1) = 0.16,$$

where $\text{Erf}(x)$ is the error function, defined as

$$\text{Erf}(x) = \frac{2}{\sqrt{\pi}} \int_0^x e^{-y^2} dy.$$

Thus, there is about a 16% chance that the oscillator will be found in the classically forbidden region.

Tutorial Problems

Problem 15.1 Use the operator approach developed in the chapter to prove that the nth harmonic oscillator energy eigenfunction obeys the following uncertainty relation

$$\delta x \delta p = \frac{\hbar}{2}(2n+1),$$

where $\delta x = \sqrt{\langle \hat{x}^2 \rangle - \langle \hat{x} \rangle^2}$ and $\delta p_x = \sqrt{\langle \hat{p}_x^2 \rangle - \langle \hat{p}_x \rangle^2}$ are fluctuations of the position and momentum operators, respectively.

Problem 15.2 Given that $\hat{a}|n\rangle = \sqrt{n}|n-1\rangle$, show that n must be a positive integer.

Problem 15.3 (a) Using the commutation relation for the position $\hat{x}$ and momentum $\hat{p} \equiv \hat{p}_x$ operators

$$[\hat{x}, \hat{p}] = i\hbar,$$

show that the annihilation and creation operators $\hat{a}$, $\hat{a}^\dagger$ of a one-dimensional harmonic oscillator satisfy the commutation relation

$$[\hat{a}, \hat{a}^\dagger] = \hat{1}.$$

(b) Show that the Hamiltonian of the harmonic oscillator

$$\hat{H} = \frac{1}{2m}\hat{p}^2 + \frac{1}{2}m\omega^2\hat{x}^2$$

can be written as

$$\hat{H} = \hbar\omega\left(\hat{a}^\dagger\hat{a} + \frac{1}{2}\right).$$

(c) Calculate the value of the uncertainty product $\Delta x \Delta p$ for a one-dimensional harmonic oscillator in the ground state $|\phi_0\rangle$, where $\Delta x = \sqrt{\langle \hat{x}^2 \rangle - \langle \hat{x} \rangle^2}$ and $\Delta p = \sqrt{\langle \hat{p}^2 \rangle - \langle \hat{p} \rangle^2}$.

Problem 15.4 Prove, by induction, the following commutation relation:

$$\left[\hat{a}, \left(\hat{a}^\dagger\right)^n\right] = n\left(\hat{a}^\dagger\right)^{n-1}.$$

Problem 15.5 *Generation of an nth wave function from the ground state wave function*
Using the normalized energy eigenfunctions of the harmonic oscillator,

$$|\phi_n\rangle = \frac{1}{\sqrt{n!}} \left(\hat{a}^\dagger\right)^n |\phi_0\rangle \, ,$$

show that

$$\hat{a}^\dagger |\phi_n\rangle = \sqrt{n+1} \, |\phi_{n+1}\rangle \, ,$$
$$\hat{a} |\phi_n\rangle = \sqrt{n} \, |\phi_{n-1}\rangle \, .$$

Problem 15.6 *Matrix representation of the annihilation and creation operators*

Write the matrix representations of the operators $\hat{a}$ and $\hat{a}^\dagger$ in the basis of the energy eigenstates $|\phi_n\rangle$, and using this representation verify the commutation relation $\left[\hat{a}, \hat{a}^\dagger\right] = \hat{1}$, where $\hat{1}$ is the unit matrix.

Problem 15.7 Introducing a dimensionless parameter $\xi = \sqrt{\frac{m\omega}{\hbar}}x$, show that:

(a) The operators $\hat{a}$ and $\hat{a}^\dagger$ can be written as

$$\hat{a} = \frac{1}{\sqrt{2}} \left(\xi + \frac{\partial}{\partial \xi}\right) \, ,$$
$$\hat{a}^\dagger = \frac{1}{\sqrt{2}} \left(\xi - \frac{\partial}{\partial \xi}\right) \, .$$

(b) The time-independent Schrödinger equation becomes

$$\frac{\partial^2 \phi}{\partial \xi^2} + \left(\frac{2E}{\hbar\omega} - \xi^2\right) \phi = 0 \, .$$

(c) Show that the wave function $\phi_1(x)$ of the $n = 1$ energy state can be written as

$$\phi_1(x) = 2\xi \, A_1 e^{-\xi^2/2} \, .$$

(d) Find the normalization constant A_1.
(e) Using (a) as the representation of the operators $\hat{a}$ and $\hat{a}^\dagger$, verify the commutation relation $\left[\hat{a}, \hat{a}^\dagger\right] = 1$.

Problem 15.8 Calculate the expectation value $\langle \hat{x} \rangle$ and the variance (fluctuations) $\sigma = \langle \hat{x}^2 \rangle - \langle \hat{x} \rangle^2$ of the position operator of a one-dimensional harmonic oscillator being in the ground state $\phi_0(x)$, using

(a) Integral definition of the average
(b) Dirac notation, which allows to express $\hat{x}$ in terms of $\hat{a}$, $\hat{a}^\dagger$, and to apply the result of question 3.
(c) Show that the average values of the kinetic and potential energies of a one-dimensional harmonic oscillator in an energy eigenstate $|\phi_n\rangle$ are equal.

Problem 15.9 Show that the nonzero minimum energy of the quantum harmonic oscillator, $E \geq \hbar\omega/2$, is a consequence of the uncertainty relation between the position and momentum operators of the oscillator.

Hint: Use the uncertainty relation for the position and the momentum operators in the state $n = 0$ and plug it into the average energy of the oscillator. Then, find the minimum of the energy in respect to δx.

Challenging Problem

Show that the probability of finding the harmonic oscillator beyond the classical turning points $x = \pm x_0$ decreases with increasing n. This example shows that the classical and quantum pictures become less and less marked with increasing n, in agreement with the correspondence principle.

Chapter 16

Quantum Theory of Hydrogen Atom

In Chapter 5, we have seen how Bohr explained the experimentally observed discrete atomic spectra. He postulated that angular momentum of the electron in a hydrogen atom is quantized, i.e.,

$$L = n\hbar, \qquad \text{where} \qquad n = 1, 2, 3, \dots . \qquad (16.1)$$

However, a careful analysis of the observed spectra showed that some spectral lines are not singlets but are composed with a few superimposed lines. Thus, the angular momentum cannot be $n\hbar$, but rather $\sqrt{l(l+1)}$, where $l = 0, 1, 2, \dots, n-1$.

It follows from the Bohr postulate that energy and also electron's orbits are quantized, that the electron can be only at some particular distances from the nucleus. In other words, the electron is not allowed to be at distances different from that predicted by the angular momentum quantization. A question then arises: where really is the electron when it makes a transition from one orbit to another?

Bohr had discussions with Schrödinger on this point, apparently without agreement. The heat of these discussions is captured in the famous statement by Schrödinger that "If all this damned quantum jumps were really to stay, I should be sorry I ever got involved with quantum theory," after which Bohr reportedly replied, "But we

Quantum Physics for Beginners
Zbigniew Ficek
Copyright © 2016 Pan Stanford Publishing Pte. Ltd.
ISBN 978-981-4669-38-2 (Hardcover), 978-981-4669-39-9 (eBook)
www.panstanford.com

others are very grateful to you that you did, since your work did so much to promote the theory".

In this chapter, we will consider Schrödinger's model of hydrogen atom and will analyze in details the quantum wave mechanics approach to the motion of the electron in the hydrogen atom. In this approach, rather than wondering about the position and motion of the electron, we will classify the electron in terms of the amount of energy that the electron has, and will represent the electron by a wave function $\Psi(\vec{r})$, which satisfies the stationary Schrödinger equation

$$\hat{H}\,\Psi(\vec{r}) = E\,\Psi(\vec{r}), \qquad (16.2)$$

where the Hamiltonian is given by

$$\hat{H} = -\frac{\hbar^2}{2m_e}\nabla^2 + \hat{V}(r), \qquad (16.3)$$

with the Coulomb potential energy of the electron

$$\hat{V}(r) = -\frac{e^2}{4\pi\,\varepsilon_0}\frac{1}{r}. \qquad (16.4)$$

Thus, the potential is spherically symmetric; it depends only on the distance r of the moving electron from the nucleus (central force).

16.1 Schrödinger Equation in Spherical Coordinates

The motion of the electron in the hydrogen atom is a three-dimensional problem, and by solving Eq. (16.2), we will find the explicit form of the wave function of the electron and its energy E.

The stationary Schrödinger equation (16.2) is not easy to solve as it stands, and the whole problem looks rather complicated when expressed in terms of Cartesian coordinates. The problem is that the differential equation (16.2) involves three variables, x, y, z, and cannot be split into a set of three separate differential equations, each involving only one variable.

Since the potential $V(r)$ has a spherical symmetry, the whole problem becomes considerably easier to solve if we work in the spherical coordinates. In the spherical coordinates, shown in

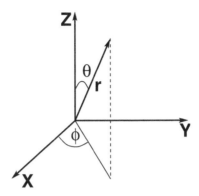

Figure 16.1 The relation between the Cartesian coordinates (x, y, z) of the position vector $\vec{r}$ and its polar coordinates (r, ϕ, θ), where r is the radial coordinate, ϕ is the azimuthal angle, and θ is the polar angle.

Fig. 16.1, we express all the coordinate variables in terms of the polar variables r, ϕ, θ. Thus, the operator ∇^2, appearing in the Schrödinger equation, when written in terms of the polar variables, takes the form

$$\nabla^2 = \frac{1}{r^2}\frac{\partial}{\partial r}\left(r^2\frac{\partial}{\partial r}\right) + \frac{1}{r^2\sin\theta}\frac{\partial}{\partial\theta}\left(\sin\theta\frac{\partial}{\partial\theta}\right) + \frac{1}{r^2\sin^2\theta}\frac{\partial^2}{\partial\phi^2}, \tag{16.5}$$

and then the Schrödinger equation can be written as

$$\frac{\partial}{\partial r}\left(r^2\frac{\partial\Psi}{\partial r}\right) + \frac{2m_e}{\hbar^2}r^2\left(E - V(r)\right)\Psi$$

$$+\frac{1}{\sin\theta}\frac{\partial}{\partial\theta}\left(\sin\theta\frac{\partial\Psi}{\partial\theta}\right) + \frac{1}{\sin^2\theta}\frac{\partial^2\Psi}{\partial\phi^2} = 0. \tag{16.6}$$

Equation (16.6) has two separate parts: the first part depends only on the distance r, whereas the second part depends only on the polar angle θ and the azimuthal angle ϕ. Thus, the wave function is of the separable form

$$\Psi(\vec{r}) = R(r)Y(\theta, \phi), \tag{16.7}$$

where $R(r)$ is called the radial part of the wave function, and $Y(\theta, \phi)$ is the angular part.

Hence, substituting Eq. (16.7) into Eq. (16.6) and dividing both sides by $R(r)Y(\theta, \phi)$, we obtain

$$\left[\frac{1}{R}\frac{d}{dr}\left(r^2 \frac{dR}{dr} \right) + \frac{2m_e r^2}{\hbar^2}(E - V(r)) \right]$$

$$= -\frac{1}{Y}\left[\frac{1}{\sin\theta}\frac{\partial}{\partial\theta}\left(\sin\theta \frac{\partial Y}{\partial\theta} \right) + \frac{1}{\sin^2\theta}\frac{\partial^2 Y}{\partial\phi^2} \right]. \quad (16.8)$$

Both sides of Eq. (16.8) depend on different variables, and thus are independent of each other and therefore must be equal to the same constant, say $-\alpha$. Consequently, we are able to separate the differential equation into two independent equations: one depending solely on r and the other on θ and ϕ:

$$\frac{1}{r^2}\frac{d}{dr}\left(r^2 \frac{dR}{dr} \right) + \frac{2m_e}{\hbar^2}(E - V(r))R + \frac{\alpha}{r^2}R = 0, \quad (16.9)$$

$$\frac{1}{\sin\theta}\frac{\partial}{\partial\theta}\left(\sin\theta \frac{\partial Y}{\partial\theta} \right) + \frac{1}{\sin^2\theta}\frac{\partial^2 Y}{\partial\phi^2} - \alpha Y = 0. \quad (16.10)$$

We will solve the above differential equations separately. First, let us consider the angular part, Eq. (16.10), that depends on the angular variables θ, ϕ.

16.1.1 Angular Part of the Wave Function: Angular Momentum

We will first show that Eq. (16.10) is, in fact, the eigenvalue equation for the square of the angular momentum operator

$$\hat{L} = \hat{r} \times \hat{p} = -i\hbar\hat{r} \times \nabla. \quad (16.11)$$

To show this, we rewrite Eq. (16.10) in the form

$$\left\{ \frac{1}{\sin\theta}\frac{\partial}{\partial\theta}\left(\sin\theta \frac{\partial}{\partial\theta} \right) + \frac{1}{\sin^2\theta}\frac{\partial^2}{\partial\phi^2} \right\}Y = \alpha Y, \quad (16.12)$$

which evidently is in the form of an eigenvalue equation of the operator

$$\hat{O} = \frac{1}{\sin\theta}\frac{\partial}{\partial\theta}\left(\sin\theta \frac{\partial}{\partial\theta} \right) + \frac{1}{\sin^2\theta}\frac{\partial^2}{\partial\phi^2}. \quad (16.13)$$

On the other hand, if we write the square of the angular momentum operator in the spherical coordinates

$$\hat{L}^2 = \hat{L} \cdot \hat{L}, \quad (16.14)$$

we find that in the operator $\hat{L}^2$ is of the form

$$\hat{L}^2 = -\hbar^2 \left\{ \frac{1}{\sin\theta} \frac{\partial}{\partial\theta} \left(\sin\theta \frac{\partial Y}{\partial\theta} \right) + \frac{1}{\sin^2\theta} \frac{\partial^2 Y}{\partial\phi^2} \right\} = -\hbar^2 \hat{O}. \quad (16.15)$$

The proof of this relation is left as an exercise for the readers.

Since the eigenvalue equation for $\hat{L}^2$ can be written as

$$\hat{L}^2 Y(\theta, \phi) = \lambda Y(\theta, \phi), \quad (16.16)$$

it then follows that $\alpha = -\lambda/\hbar^2$, where λ is the eigenvalue of $\hat{L}^2$.

We now return to Eq. (16.10), which we can write as

$$\sin\theta \frac{\partial}{\partial\theta} \left(\sin\theta \frac{\partial Y}{\partial\theta} \right) - \alpha \sin^2\theta Y + \frac{\partial^2 Y}{\partial\phi^2} = 0. \quad (16.17)$$

This equation contains two separate parts: one dependent only on θ and the other dependent only on ϕ. Therefore, the solution to Eq. (16.17) will be of the form

$$Y(\theta, \phi) = X(\theta)\Phi(\phi). \quad (16.18)$$

Hence, substituting Eq. (16.18) into Eq. (16.17) and dividing both sides by $X(\theta)\Phi(\phi)$, we obtain

$$\frac{1}{X} \sin\theta \frac{d}{d\theta} \left(\sin\theta \frac{dX}{d\theta} \right) - \alpha \sin^2\theta = -\frac{1}{\Phi} \frac{d^2\Phi}{d\phi^2}, \quad (16.19)$$

where $X \equiv X(\theta)$ and $\Phi \equiv \Phi(\phi)$.

As before, both sides must be equal to a constant, say m^2. Thus

$$\frac{1}{X} \sin\theta \frac{d}{d\theta} \left(\sin\theta \frac{dX}{d\theta} \right) - \alpha \sin^2\theta = m^2, \quad (16.20)$$

$$\frac{1}{\Phi} \frac{d^2\Phi}{d\phi^2} = -m^2. \quad (16.21)$$

Hence, we have separated Eq. (16.17) into two differential equations, that we can solve separately.

16.1.2 *Azimuthal Part of the Wave Function*

First, we will solve the equation for the azimuthal part of the wave function, which is particularly simple. We rewrite Eq. (16.21) in the form

$$\frac{d^2\Phi}{d\phi^2} = -m^2\Phi, \quad (16.22)$$

and readily find that a general solution to Eq. (16.22) is of the form

$$\Phi(\phi) = A \exp(im\phi) , \tag{16.23}$$

where A is a constant.

Since in rotation, ϕ and $\phi + 2\pi$ correspond to the same position in space, we have $\Phi(\phi) = \Phi(\phi + 2\pi)$. Thus

$$\exp(im\phi) = \exp[im(\phi + 2\pi)] , \tag{16.24}$$

from which we find that

$$\exp(i2\pi m) = 1 . \tag{16.25}$$

However, this condition is satisfied *only* when m is an integer, $m = 0, \pm1, \pm2, \ldots$.

Hence, the constant m^2 is not an arbitrary number and is an integer.

Using the normalization condition

$$1 = \int_0^{2\pi} |\Phi(\phi)|^2 \, d\phi = 2\pi |A|^2 , \tag{16.26}$$

we can write the final form of the azimuthal part of the wave function $\Phi(\phi)$ as

$$\Phi_m(\phi) = \frac{1}{\sqrt{2\pi}} \exp(im\phi) , \tag{16.27}$$

where m is an integer, and the subscript m has been introduced to indicate the dependence of the wave function upon the quantum number m.

16.1.3 *Polar Component of the Wave Function*

The next step in the solution is to find $X(\theta)$, the polar component of the wave function that is a solution to the differential equation (16.20). The solution is rather complicated and is given in terms of special functions.

The procedure of solving the differential equation (16.20) is as follows: If we multiply both sides of the equation by X and divide by $\sin^2 \theta$, we obtain after a slight rearrangement

$$\frac{1}{\sin\theta} \frac{d}{d\theta} \left(\sin\theta \frac{dX}{d\theta} \right) - \left(\alpha + \frac{m^2}{\sin^2 \theta} \right) X = 0 . \tag{16.28}$$

Introducing a new variable $z = \cos\theta$, and noting that

$$\frac{d}{d\theta} = -\sqrt{1-z^2}\,\frac{d}{dz}\,, \tag{16.29}$$

we find that Eq. (16.28) becomes

$$(1-z^2)\frac{d^2X}{dz^2} - 2z\frac{dX}{dz} - \left(\alpha + \frac{m^2}{1-z^2}\right)X = 0\,, \tag{16.30}$$

which after a slight rearrangement can be written as

$$\frac{d}{dz}\left[(1-z^2)\frac{dX}{dz}\right] - \left(\alpha + \frac{m^2}{1-z^2}\right)X = 0\,. \tag{16.31}$$

This differential equation is known in mathematics as the *generalized Legendre* differential equation, and its solutions are the *associated Legendre polynomials*. For $m = 0$, the equation is called the *ordinary Legendre* differential equation whose solution is given by the *Legendre polynomials*.

Equation (16.31) has singularities (poles) at $z = \pm 1$. However, the desired solution should be single valued, finite, and continuous on the interval $-1 \le z \le 1$ to represent the wave function of the electron.

To find the physically acceptable solution to this equation, we will check what solution could be continuous near the poles.

We first find a possible finite solution near $z = 1$. Substituting $x = 1 - z$, we have $dx = -dz$, and then in terms of x, the differential equation (16.31) takes the form

$$\frac{d}{dx}\left[x(2-x)\frac{dX}{dx}\right] - \left(\alpha + \frac{m^2}{x(2-x)}\right)X = 0\,. \tag{16.32}$$

The standard procedure for solving differential equations like Eq. (16.32) is to assume that $X(x)$ can be given in terms of a power series in x:

$$X(x) = x^s \sum_{j=0}^{\infty} a_j x^j\,. \tag{16.33}$$

Substituting this into the differential equation for X, we get

$$2s^2 a_0 x^{s-1} + (s+1)(2sa_1 - sa_0 + 2a_1)x^s + \ldots$$
$$- \left(\alpha + \frac{m^2}{x(2-x)}\right)(a_0 + a_1 x + \ldots)x^s = 0\,. \tag{16.34}$$

Near $x \approx 0$, we can replace $x(2 - x)$ by $2x$ and obtain

$$\left(2s^2 a_0 - \frac{m^2}{2} a_0\right) x^{s-1} + (\ldots)x^s \ldots = 0 . \tag{16.35}$$

This equation is satisfied for all x only if the coefficients at x^s, $x^{s\pm1}, \ldots$ are zero. From this, we find that

$$s = \pm\frac{1}{2}|m| . \tag{16.36}$$

We take only $s = +\frac{1}{2}|m|$ as for $s = -\frac{1}{2}|m|$ the solution to $X(x)$ at $x = 0$ would go to infinity. We require the solution to the wave function to be finite at any point x.

Thus, the solution that is continuous near $x = 0$ is of the form

$$X(x) = x^{\frac{1}{2}|m|} \sum_{j=0}^{\infty} a_j x^j \tag{16.37}$$

or in terms of z

$$X(z) = (1 - z)^{\frac{1}{2}|m|} \sum_{j=0}^{\infty} a_j' z^j . \tag{16.38}$$

Using the same procedure, we can show that near the pole $z = -1$, the continuous solution is

$$X(z) = (1 + z)^{\frac{1}{2}|m|} \sum_{j=0}^{\infty} a_j'' z^j . \tag{16.39}$$

Hence, we will try to find the solution to Eq. (16.31) in the form

$$X(z) = \left(1 - z^2\right)^{\frac{1}{2}|m|} \sum_{j=0}^{\infty} b_j z^j . \tag{16.40}$$

Substituting this equation into the differential equation for $X(z)$ and collecting all terms at the same powers of z^j, we obtain

$$\sum_{j} \{(j + 1)(j + 2)b_{j+2} - j(j - 1)b_j$$
$$- 2(|m| + 1)jb_j - (\alpha + |m| + m^2)b_j\} z^j = 0 . \tag{16.41}$$

Hence, we get a recurrence relation for the coefficients b_j

$$b_{j+2} = \frac{(j + |m|)(j + |m| + 1) + \alpha}{(j + 1)(j + 2)} b_j . \tag{16.42}$$

Thus, if we know the first two coefficients, b_0 and b_1, then we can determine the whole series. However, there is a problem with the series. Since $b_{j+2} > b_j$, the series diverges (logarithmically) for $z = \pm 1$, which means that $X(z)$ is not finite at $z = \pm 1$. But the wave function must be finite everywhere, including the polar directions $z = \pm 1$.

There is a simple way out of this dilemma.

If the series representing $X(z)$ terminates at a certain b_{j_0}, so that the coefficients b_j are zero for $j > j_0$, the wave function $X(z)$ will be finite everywhere. In other words, if $X(z)$ is a polynomial with a finite number of terms instead of an infinite series, it is acceptable.

Therefore, we terminate the series at some $j = j_0$, i.e., we assume that $b_{j_0+1} = b_{j_0+2} = \ldots = 0$. The series terminating at $j = j_0$ indicates that

$$(j_0 + |m|)(j_0 + |m| + 1) + \alpha = 0 . \tag{16.43}$$

Introducing

$$l = j_0 + |m| , \tag{16.44}$$

we see that $l \geq |m|$, and

$$\alpha = -l(l+1) , \qquad l = 0, 1, 2, \ldots \tag{16.45}$$

Hence, we see that the eigenvalues of the angular momentum are quantized

$$\hat{L}^2 : \lambda = \hbar^2 l(l+1) ,$$
$$\hat{L} : \lambda = \hbar \sqrt{l(l+1)} . \tag{16.46}$$

The integer number l is called the *angular momentum quantum number*. Since $l \geq |m|$, the number m is limited to absolute values not larger than l.

After the termination of the series, we get the solution to the wave function $X(z)$, which is in the form of polynomials, called the associated Legendre polynomials

$$X_{lm}(z) = (1-z)^{\frac{1}{2}|m|} \sum_{j=0}^{l-|m|} b_j z^j , \tag{16.47}$$

where the subscript lm has been introduced to indicate the dependence of the wave function on the quantum numbers l and m.

Important: In the sum over j, we take either even or odd terms in z^j depending on whether $l - |m|$ is even or odd.

For example, when $l - |m|$ is even, we take $b_0 \neq 0$ and put $b_1 = 0$, so that the solution is given in terms of only even j. For $l - |m|$ odd, we put $b_0 = 0$ and take $b_1 \neq 0$, so that the solution is given in terms of odd j.

Why we cannot accept both the even and odd solutions at the same time?

The answer is as follows: We cannot accept both the even and odd solutions at the same time because in this case the solution $X(z)$ would not be a single-valued function.

For example, for $b_0 \neq 0$, we have $\alpha = -|m| - m^2$, but for $b_1 \neq 0$, we have $\alpha = -2 - 3|m| - m^2$. If we would accept both the solutions at the same time, the wave function would have two different values.

As an illustration: The first few associated Legendre polynomials are

$$X_{00}(z) = b_0 \,,$$
$$X_{10}(z) = b_1 z \,,$$
$$X_{11}(z) = b_0 \sqrt{1 - z^2} \,, \tag{16.48}$$

where the coefficients $b_0, b_1, \ldots$ are found from the normalization of the wave functions $X_{lm}(z)$.

The first few normalized complete angular parts of the wave function of the electron $Y_{lm}(\theta, \phi) = X_{lm}(\theta)\Phi_m(\phi)$, the so-called *spherical harmonics*, are:

$$Y_{00}(\theta, \phi) = \frac{1}{\sqrt{4\pi}} \,,$$
$$Y_{10}(\theta, \phi) = \sqrt{\frac{3}{4\pi}} \cos\theta \,,$$
$$Y_{11}(\theta, \phi) = -\sqrt{\frac{3}{8\pi}} \sin\theta \, e^{i\phi} \,,$$
$$Y_{1-1}(\theta, \phi) = \sqrt{\frac{3}{8\pi}} \sin\theta \, e^{-i\phi} \,. \tag{16.49}$$

All the spherical harmonics have basically the same mathematical structure and are in the form of powers of the sine and cosine functions.

16.1.4 *Physical Interpretation of the Quantum Number m*

Before proceeding further with the procedure of finding the complete wave function of the electron, let us discuss the physical meaning of the quantum number m. We have already shown that the azimuthal part of the wave function is given by

$$\Phi_m(\phi) = \frac{1}{\sqrt{2\pi}} \exp(im\phi), \qquad m = 0, \pm 1, \pm 2, \ldots, \pm l. \quad (16.50)$$

Consider the z-component, $\hat{L}_z$, of the angular momentum. We will try to find the eigenvalues and eigenfunctions of $\hat{L}_z$:

$$\hat{L}_z \Phi = \mu \Phi. \quad (16.51)$$

It is convenient to write the operator $\hat{L}_z$ in the spherical coordinates, where it takes the form

$$\hat{L}_z = -i\hbar \frac{\partial}{\partial \phi}. \quad (16.52)$$

Then, we get from Eq. (16.51) a simple differential equation

$$-i\hbar \frac{\partial \Phi}{\partial \phi} = \mu \Phi, \quad (16.53)$$

whose solution is

$$\Phi(\phi) = A \exp\left(\frac{i}{\hbar} \mu \phi\right), \quad (16.54)$$

where A is a constant.

Using the same argument as before that in rotation ϕ and $\phi + 2\pi$ correspond to the same position in space, we find that

$$\mu = m\hbar, \qquad m = 0, \pm 1, \pm 2, \ldots \quad (16.55)$$

Thus, the azimuthal part of the wave function is the wave function of the z-component of the angular momentum, and the number m is the z-component angular momentum quantum number.

Note that the component $\hat{L}_z$, whose eigenvalue is $m\hbar$, can never be as long as the vector $\hat{L}$, whose magnitude is $\hbar\sqrt{l(l+1)}$. This is illustrated in the following example.

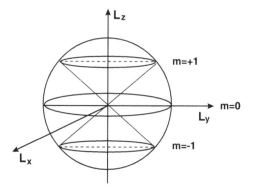

Figure 16.2 Angular momentum quantization for $l = 1$.

Example

Consider angular momentum with $l = 1$. In this case, the eigenvalue of $\hat{L}$ is $\sqrt{2}\hbar$, and $\hat{L}_z$ can have three values $+\hbar$, 0, $-\hbar$. Thus, the orientation of $\hat{L}$ along the z-axis is quantized. The vector $\hat{L}$ processes around the z-axis, sweeping out cones of revolution around that axis. This is shown in Fig. 16.2. The quantization of the orientation of $\hat{L}$ along its z-axis is called *space quantization*.

16.2 Radial Part of the Wave Function

The final step in the procedure of finding the wave function of the electron is to determine the remaining radial part R of the wave function (Eq. (16.9)).

We start with a simplification of Eq. (16.9) by introducing new variables

$$\beta^2 = -\frac{2m_e E}{\hbar^2}, \quad \lambda = \frac{m_e e^2}{4\pi \varepsilon_0 \hbar^2 \beta}, \quad \rho = 2\beta r, \quad (16.56)$$

and substituting the explicit form for $V(r)$ (Eq. (16.4)), and $\alpha = -l(l+1)$.

After this simplification, the differential equation (16.9) takes the form

$$\frac{1}{\rho^2}\frac{d}{d\rho}\left(\rho^2\frac{dR}{d\rho}\right) + \left[\frac{\lambda}{\rho} - \frac{1}{4} - \frac{l(l+1)}{\rho^2}\right]R = 0. \qquad (16.57)$$

We begin the solution to Eq. (16.57) by finding the asymptotic form that R must have if we want $R \to 0$ as $r \to \infty$.

In the limit of $\rho \gg 1$, we can ignore in Eq. (16.57) the terms proportional to $1/\rho$ and $1/\rho^2$ and obtain an asymptotic equation

$$\frac{d^2 R}{d\rho^2} - \frac{1}{4}R = 0. \qquad (16.58)$$

It is a simple second-order differential equation whose solution is of the exponential form

$$R(\rho) = Ae^{-\frac{1}{2}\rho} + Be^{\frac{1}{2}\rho}, \qquad (16.59)$$

where A and B are constants.

In the limit of $r \to \infty$ ($\rho \to \infty$), the second term in Eq. (16.59) goes to infinity and then the wave function R would be infinite. Since the wave function must be finite for all r, we ignore the second term, leaving only the first term in Eq. (16.59) as an acceptable asymptotic solution.

Following this asymptotic behavior, we expect the general solution to Eq. (16.57) in the form

$$R(\rho) = e^{-\frac{1}{2}\rho}F(\rho), \qquad (16.60)$$

where $F(\rho)$ is a function of ρ that remains to be found.

To find $F(\rho)$, we substitute Eq. (16.60) into (16.57) and obtain

$$\frac{d^2 F}{d\rho^2} + \left(\frac{2}{\rho} - 1\right)\frac{dF}{d\rho} + \left[\frac{\lambda}{\rho} - \frac{1}{\rho} - \frac{l(l+1)}{\rho^2}\right]F = 0. \qquad (16.61)$$

This equation contains several terms, which become infinity at $\rho = 0$. Since the solution must be finite everywhere including $r = 0$, we will look for a solution in the form of a power series in ρ:

$$F(\rho) = \rho^s L(\rho), \qquad (16.62)$$

where s and $L(\rho)$ have to be determined.

Substituting Eq. (16.62) into Eq. (16.61), we find

$$\rho^{s+2}\frac{d^2 L}{d\rho^2} + 2s\rho^{s+1}\frac{dL}{d\rho} + s(s-1)\rho^s L + 2\rho^{s+1}\frac{dL}{d\rho} + 2s\rho^s L$$

$$-\rho^{s+2}\frac{dL}{d\rho} - s\rho^{s+1}L + (\lambda - 1)\rho^{s+1}L - l(l+1)\rho^s L = 0. \qquad (16.63)$$

In order for this equation to hold for all values of ρ, the coefficients at all powers of ρ must be zero. In particular, the coefficient at ρ^s is

$$\{s(s+1) - l(l+1)\} L, \tag{16.64}$$

and vanishes when

$$s = l \quad \text{or} \quad s = -(l+1). \tag{16.65}$$

For $s = -(l+1)$, the function $F(\rho) = \rho^{-(l+1)} L(\rho)$ diverges as $\rho \to \infty$. Hence, we ignore this solution leaving $s = l$ as the only acceptable solution to $F(\rho)$.

We have already determined s; what left is to determine $L(\rho)$. The function $L(\rho)$ is found from Eq. (16.63). When we substitute $s = l$ into Eq. (16.63) and divide both sides by ρ^{l+1}, we obtain a second-order differential equation

$$\rho \frac{d^2 L}{d\rho^2} + [2(l+1) - \rho] \frac{dL}{d\rho} + (\lambda - l - 1) L = 0. \tag{16.66}$$

The standard procedure for solving differential equations like Eq. (16.66) is to assume that $L(\rho)$ can be expanded in a power series in ρ:

$$L(\rho) = \sum_{j=0}^{\infty} b_j \rho^j, \tag{16.67}$$

and then to determine the values of the coefficients b_j.

Substituting Eq. (16.67) into Eq. (16.66), we obtain the recursion relation for the coefficients b_j:

$$b_{j+1} = \frac{(j+l+1-\lambda)}{2(j+1)(l+1) + j(j+1)} b_j, \tag{16.68}$$

which enables us to find the coefficients $b_1, b_2, b_3, \ldots$ in terms of b_0.

We now check whether the coefficients converge as $j \to \infty$.

For a large j ($j \gg 1$), we get

$$\frac{b_{j+1}}{b_j} \approx \frac{1}{j}, \tag{16.69}$$

which shows that the coefficients converge as $j \to \infty$.

Because $L(\rho)$ is multiplied by the exponential function $\rho^l e^{-\frac{1}{2}\rho}$, the wave function R will vanish at $\rho \to \infty$ only if

$$\lim_{\rho \to \infty} L(\rho) < e^{\frac{1}{2}\rho}. \tag{16.70}$$

However, the function $L(\rho)$ behaves at $\rho \to \infty$ as the exponent function e^ρ. To show this, expand e^ρ into a series

$$e^\rho = 1 + \rho + \frac{\rho^2}{2!} + \frac{\rho^3}{3!} + \dots + \frac{\rho^j}{j!} + \frac{\rho^{j+1}}{(j+1)!} \dots \quad (16.71)$$

The ratio of the coefficients at ρ^{j+1} and ρ^j is equal to

$$\frac{j!}{(j+1)!} = \frac{1}{j+1} \approx \frac{1}{j} \qquad \text{for} \quad j \to \infty. \quad (16.72)$$

Hence, for large j, the series $L(\rho)$ behaves as e^ρ. Therefore, the radial function

$$R(\rho) = e^{-\frac{1}{2}\rho} \rho^l L(\rho) \quad (16.73)$$

would behave as $\rho^l e^{\frac{1}{2}\rho}$, which does not vanish as $\rho \to \infty$. This means that the wave function $R(\rho)$ would not be a physically acceptable wave function.

As before, we solve this dilemma in the following way. If the series representing $R(\rho)$ terminates at a certain b_{j_0}, so that all coefficients b_j are zero for $j > j_0$, the wave function $R(\rho)$ will go to zero as $r \to \infty$ because of the exponential factor $e^{-\frac{1}{2}\rho}$.

Therefore, we terminate the series at some $j = j_0$, which according to Eq. (16.68) corresponds to $j_0 = \lambda - l - 1$. In other words, the condition $j_0 = \lambda - l - 1$ is a necessary and sufficient condition for the wave function R to be continuous for all r and vanish as $r \to \infty$.

Next, denoting $j_0 + l + 1 = n$, we have $\lambda = n > 0$, i.e., $n = 1, 2, 3, \dots$. In other words, n can never be zero. Moreover, we see that $n > l$. We call the integer number n — the *principal quantum number*.

We have found that $\lambda (= n)$ is a discrete number, so β too is a discrete number, and from that we find the energy of the electron

$$E = -\frac{1}{(4\pi\varepsilon_0)^2} \frac{m_e e^4}{2\hbar^2} \frac{1}{n^2}. \quad (16.74)$$

We can introduce a constant

$$a_o = \frac{4\pi\varepsilon_0\hbar^2}{m_e e^2}, \quad (16.75)$$

called the *Bohr radius*, and then

$$E = -\frac{1}{4\pi\varepsilon_0} \frac{e^2}{2a_o n^2}. \quad (16.76)$$

Thus, the energy of the electron in the hydrogen atom is quantized.[a] Notice that Eq. (16.76) agrees perfectly with the prediction of the Bohr theory of the hydrogen atom (see Eq. (5.11)). Hence, we see that Bohr's concept that the electron can exist only in discrete energy levels survived the transition to quantum wave mechanics.

Since $\rho = 2\beta r$, and $\beta = 1/(a_o n)$, the radial part of the wave function can be written as

$$R_{nl}(r) = e^{-\beta r} (2\beta r)^l \, L_n^l(r) \,, \tag{16.77}$$

where

$$L_n^l(r) = \sum_{j=0}^{n-l-1} b_j \, (2\beta r)^j \tag{16.78}$$

are the *associated Laquerre polynomials* of order $(n - l - 1)$.

The coefficients b_j are found from the normalization of the radial function

$$\int_0^\infty dr r^2 |R_{nl}(r)|^2 = 1 \,. \tag{16.79}$$

Once the radial part of the wave function is known, the solution to the problem of the hydrogen atom is completed by writing down the normalized wave function of the electron

$$\Psi_{nlm}(r, \theta, \phi) = R_{nl}(r) Y_{lm}(\theta, \phi) \,, \tag{16.80}$$

where the subscript *nlm* indicates the dependence of the wave function on the quantum numbers n, l, and m. Thus, for each set of quantum numbers (n, l, m), there is a different wave function Ψ_{nlm}.

Summary

The eigenvalues of the energy of the electron in the hydrogen atom are quantized

$$E_n = -\frac{1}{4\pi\varepsilon_0} \frac{e^2}{2a_o n^2} \,, \tag{16.81}$$

---

[a] An interesting observation: The energy E depends solely on the quantum number n, not on the quantum numbers l and m. Why? I leave the answer to this question for the reader.

and the corresponding eigenfunctions are

$$\Psi_{nlm}(r, \theta, \phi) = R_{nl}(r)Y_{lm}(\theta, \phi), \qquad (16.82)$$

where the discrete (quantum) numbers are

$$n = 1, 2, 3, \ldots, \infty,$$
$$l = 0, 1, 2, \ldots, n-1,$$
$$m = 0, \pm 1, \pm 2, \ldots, \pm l. \qquad (16.83)$$

The angular momentum states are often indicated by letters $s, p, d, \ldots$, such that s corresponds to $l = 0$ state, p to $l = 1$, and so on according to the following scheme:

$$l = 0, 1, 2, 3, 4, 5, 6, \ldots,$$
$$s, p, d, f, g, h, i, \ldots \qquad (16.84)$$

This particular notation that is widely used in atomic physics originated from the classification of atomic spectra into series called (s) *sharp*, (p) *principal*, (d) *diffusive*, and (f) *fundamental*. Thus, an s state is one with no angular momentum, a p state has the angular momentum $\sqrt{2}\hbar$, etc.

The analysis of the solution to the Schrödinger equation shows how inevitably quantum numbers appear in Schrödinger's model of the hydrogen atom. Namely, the important condition for obtaining physically acceptable solution to the wave function of the electron is that n, l, and m are integer parameters. This is in contrast to Bohr's model, where the quantization of the angular momentum was postulated without any evident reasons.

Few normalized eigenfunctions of the electron

$$\Psi_{100} = \frac{1}{\sqrt{\pi a_o^3}} e^{-r/a_o},$$

$$\Psi_{200} = \frac{1}{\sqrt{8\pi a_o^3}} \left(1 - \frac{r}{2a_o}\right) e^{-r/(2a_o)},$$

$$\Psi_{210} = \frac{1}{\sqrt{32\pi a_o^3}} \frac{r}{a_o} e^{-r/(2a_o)} \cos\theta. \qquad (16.85)$$

Note that eigenfunctions with $l = 0$ have spherical symmetry, i.e., are independent of the angular variables θ and ϕ.

The absolute square of the wave function $|\Psi_{nlm}(r, \theta, \phi)|^2$ is the probability density of finding the electron at the point $\vec{r}(r, \theta, \phi)$, and

$$P_{nlm} = |\Psi_{nlm}(r, \theta, \phi)|^2 dV = 4\pi r^2 |\Psi_{nlm}(r, \theta, \phi)|^2 dr d\theta d\phi \qquad (16.86)$$

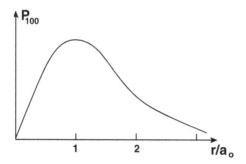

Figure 16.3   Probability function of the electron in the state $(nlm) = (100)$.

is the probability of finding the electron in a small volume $dV = dr d\theta d\phi$ around the point $\vec{r}$.

The maximum value of P_{nlm}, which is the most probable distance of the electron from the nucleus, differs from the expectation (average) distance $\langle r \rangle$, given by

$$\langle r \rangle = \int \Psi^*_{nlm} r \Psi_{nlm} dV . \tag{16.87}$$

Examples of the probability distribution P_{nlm} are plotted in Figs. 16.3 and 16.4.

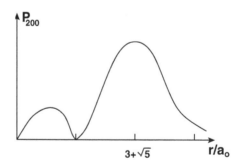

Figure 16.4   Probability function of the electron in the state $(nlm) = (200)$.

Interesting properties of the probability function P_{nlm}:

(1) For $n = 1$, the probability has one maximum exactly at $r = a_o$, the orbital radius of the first energy level in the Bohr model.

(2) For $(n = 2, \; l = 0, \; m = 0)$, the probability shows two maxima located at $r \neq n^2 a_o$.

(3) Only for states such that $n = l + 1$, the probability shows one maximum located at $r = n^2 a_o$, the orbital radius of the nth energy level in the Bohr model.

Remark on the difference between Bohr's and Schrödinger's models of the hydrogen atom: We have seen that according to Schrödinger's model of the hydrogen atom, the energy of the electron is quantized and the energy spectrum is exactly the same as that predicted by Bohr's model.

An essential difference between Bohr's and Schrödinger's models is that there are three (ignoring spin) quantum numbers, n, l, m, rather than the one number n describing a state of the electron. Moreover, in Bohr's model the atom is, at each instant, in one and only one energy state, i.e., at each instant, the state of the atom is definite, but according to Schrödinger's model, the atom can exist in a superposition of states. In other words, the atom can be simultaneously in more than one energy state at a given instant. This is clearly incompatible with Bohr's model.

Worked Example

The normalized wave function for the ground state of a hydrogen atom has the form

$$\Psi(r) = A e^{-r/a_o},$$

where $A = 1/\sqrt{\pi a_o^3}$ is a constant, $a_o = 4\pi \varepsilon_0 \hbar^2 / m e^2$ is the Bohr radius, and r is the distance between the electron and the nucleus. Show the following:

(a) The expectation value of r is $\frac{3}{2} a_o$.

(b) The most probable value of r is $r = a_o$.

Solution

(a) From the definition of expectation value, we find

$$\langle r \rangle = \int \Psi^* (r) r \Psi (r) \, dV = 4\pi A^2 \int_0^\infty r^3 e^{-2\beta r} dr \,,$$

where $\beta = 1/a_o$, and we have transformed the integral over dV into spherical coordinates with $dV = 4\pi r^2 dr$.

Performing the integration, we obtain

$$\langle r \rangle = 4\pi A^2 \frac{6}{(2\beta)^4} = \frac{24\,\pi\,A^2}{16}\frac{1}{\beta^4} = \frac{3}{2}\pi \frac{1}{\pi a_o^3} a_o^4 = \frac{3}{2}a_o \,.$$

Thus, the average distance of the electron from the nucleus in the state Ψ is $3/2$ of the Bohr radius.

(b) The most probable value of r is that where the probability of finding the electron is maximal.

Thus, we first calculate the probability of finding the electron at a point r:

$$P(r) = 4\pi r^2 |\Psi (r)|^2 = 4\pi r^2 A^2 e^{-2\beta r} = \frac{4r^2}{a_o^3} e^{-2\beta r} \,.$$

Maximum of $P(r)$ is where $dP(r)/dr = 0$. Hence

$$\frac{dP(r)}{dr} = \frac{8r}{a_o^3} e^{-2\beta r} - \frac{8\beta r^2}{a_o^3} e^{-2\beta r} \,.$$

Thus

$$\frac{dP(r)}{dr} = 0 \qquad \text{when} \qquad \beta r = 1 \,,$$

from which, we find

$$r = \frac{1}{\beta} = a_o \,.$$

Note that this result agrees with the prediction of the Bohr model that the radius of the $n = 1$ orbit is equal to a_o. The maximum of the probability at $r = a_o$ is

$$P_{\max}(a_o) = \frac{4a_o^2}{a_o^3} e^{-2} = \frac{4}{e^2 a_o} \,.$$

In summary of the solution: The expectation and most probable values of r are not the same. This is because the probability curve $P_{100}(r)$ is not symmetric about the maximum at a_o (see Fig. 16.3). Thus, values of r greater than a_o are weighted more heavily in the equation for the expectation value than values smaller than a_o. This results in the expectation value $\langle r \rangle$ exceeding a_o for this probability distribution.

Discussion Problems

Problem D7 Explain why an external (static) magnetic field affects the motion of the electron described only by the quantum number m, i.e., the azimuthal component of the motion, and leads to a splitting of the energy levels corresponding to different m (the Zeeman effect).

Problem D8 One can see from Eq. (16.49) that the solutions to the Schrödinger equation for the angular part of the wave function are not spherically symmetric except for $l = 0$. Explain, how the solutions are not spherically symmetric despite the fact that the potential $V(r)$ is spherically symmetric.

Tutorial Problems

Problem 16.1 According to the Bohr model of the hydrogen atom, the smallest distance of the electron from the nucleus is equal to the Bohr radius a_o, i.e., when the electron is in the $n = 1$ state.
What does the quantum wave mechanics say about it? Calculate the probability of finding the electron closer to the nucleus than the Bohr radius.

Problem 16.2 *Extension of the above worked example to a hydrogen-like atom with nuclear charge Ze.*
The normalized wave function of the ground state of a hydrogen-like atom with nuclear charge Ze has the form

$$|\Psi\rangle = A e^{-\beta r} ,$$

where A and β are real constants, and r is the distance between the electron and the nucleus. The Hamiltonian of the atom is given by

$$\hat{H} = -\frac{\hbar^2}{2m}\nabla^2 - \frac{Ze^2}{4\pi\varepsilon_0}\frac{1}{r}.$$

Show that

(a) $A^2 = \beta^3/\pi$.
(b) $\beta = Z/a_o$, where a_o is the Bohr radius.
(c) The energy of the electron is $E = -Z^2 E_0$,
    where $E_0 = e^2/(8\pi\varepsilon_0 a_o)$.
(d) The expectation values of the potential and kinetic energies are $2E$ and $-E$, respectively.

Problem 16.3 *Angular momentum operator*
Consider the angular momentum operator $\hat{L} = \hat{r} \times \hat{p}$. Show that:

(a) The operator $\hat{L}$ is Hermitian.
    (Hint: Show that the components L_x, L_y, L_z are Hermitian).
(b) The components of $\hat{L}$ (L_x, L_y, L_z) do not commute.
(c) The square of the angular momentum $\hat{L}^2$ commutes with each of the components L_x, L_y, L_z.
(d) In the spherical coordinates, the components and the square of the angular momentum can be expressed as

$$L_x = -i\hbar\left(-\sin\phi\frac{\partial}{\partial\theta} - \frac{\cos\phi\cos\theta}{\sin\theta}\frac{\partial}{\partial\phi}\right),$$

$$L_y = -i\hbar\left(\cos\phi\frac{\partial}{\partial\theta} - \frac{\sin\phi\cos\theta}{\sin\theta}\frac{\partial}{\partial\phi}\right),$$

$$L_z = -i\hbar\frac{\partial}{\partial\phi},$$

$$L^2 = -\hbar^2\left[\frac{1}{\sin^2\theta}\frac{\partial^2}{\partial\phi^2} + \frac{1}{\sin\theta}\frac{\partial}{\partial\theta}\left(\sin\theta\frac{\partial}{\partial\theta}\right)\right].$$

Problem 16.4 *Particle in a potential of central symmetry*
A particle of mass m moves in a potential of central symmetry, i.e., $V(x, y, z) = V(r)$. The energy of the particle is given by the

Hamiltonian

$$\hat{H} = -\frac{\hbar^2}{2m}\nabla^2 + \hat{V}(r).$$

Show that $\hat{H}$ commutes with the angular momentum $\hat{L}$ of the particle.

Problem 16.5 *Quantized motion of a rotating mass*

Suppose that a particle of mass m can rotate around a fixed point A, such that $r = $ constant, and $\theta = \pi/2 = $ constant.

(a) Show that the motion of the particle is quantized.
(b) Show that the only acceptable solutions to the wave function of the particle are those corresponding to the positive energies $(E > 0)$ of the particle.

Problem 16.6 *Transition dipole moments*

The electron in a hydrogen atom can be in two states of the form

$$\Psi_1(r) = \sqrt{2}Ne^{-r/a_o},$$

$$\Psi_2(r) = \frac{N}{4a_o}re^{-r/(2a_o)}\cos\theta,$$

where $r = (x^2 + y^2 + z^2)^{\frac{1}{2}}$, $\cos\theta = z/r$, $N = 1/\sqrt{2\pi a_o^3}$, and a_o is the Bohr radius. Using the spherical coordinates, in which

$$x = r\sin\theta\cos\phi,$$
$$y = r\sin\theta\sin\phi,$$
$$z = r\cos\theta,$$

and

$$\int dV = \int_0^\infty \int_0^\pi \int_0^{2\pi} r^2\sin\theta dr d\theta d\phi,$$

(a) Show that the functions $\Psi_1(r)$, $\Psi_2(r)$ are orthogonal.
(b) Calculate the matrix element $(\Psi_1(r), \hat{r}\Psi_2(r))$ of the position operator $\hat{r}$ between the states $\Psi_1(r)$ and $\Psi_2(r)$.
(The matrix element is related to the atomic electric dipole moment between the states $\Psi_1(r)$ and $\Psi_2(r)$, defined as $(\Psi_1(r), \hat{\mu}\Psi_2(r)) = e(\Psi_1(r), \hat{r}\Psi_2(r))$.)
(c) Show that the average values of the kinetic and potential energies in the state $\Psi_1(r)$ satisfy the relation $\langle E_k \rangle = -\frac{1}{2}\langle V \rangle$.

Problem 16.7 The wave functions of the electron in the states $n = 1$ and $n = 2, l = 1, m = 0$ of the hydrogen atom are

$$\Psi_{100} = \frac{1}{\sqrt{\pi a_o^3}} \, e^{-r/a_o} \, ,$$

$$\Psi_{210} = \frac{1}{\sqrt{32\pi a_o^3}} \, \frac{r}{a_o} \, e^{-r/(2a_o)} \cos\theta \, ,$$

where a_o is the Bohr radius.

(a) Calculate the standard deviation $\sigma^2 = \langle r^2 \rangle - \langle r \rangle^2$ of the position of the electron in these two states to determine in which of these states the electron is more stable in the position.
(b) The electron is found in a state

$$\Psi = \sqrt{\frac{8}{\pi a_o^3}} \, e^{-2r/a_o} \, .$$

Determine what is the probability that the state Ψ is the ground state $(n = 1)$ of the hydrogen atom.

Challenging Problem: *Eigenfunctions of the angular momentum*

The eigenfunctions of the angular momentum $\hat{L}$ of the electron in a hydrogen atom for $l = 1$ are

$$Y_{10}\,(\theta, \, \phi) = \sqrt{\frac{3}{4\pi}} \cos\theta \, , \qquad Y_{1\pm1}\,(\theta, \, \phi) = \mp\sqrt{\frac{3}{8\pi}} \sin\theta\, e^{\pm i\phi} \, .$$

(a) Show that the eigenfunctions are also eigenfunctions of the $\hat{L}_z$ component of the angular momentum.
(b) Show that the eigenfunctions are not eigenfunctions of the $\hat{L}_x$ component of the angular momentum.
(c) Find the matrix representation of $\hat{L}_x$ in the basis of the eigenfunctions of $\hat{L}$.
(d) Find the eigenvalues and eigenfunctions of $\hat{L}_x$ in the basis of the eigenfunctions of $\hat{L}$.

Chapter 17

Quantum Theory of Two Coupled Particles

In the preceding chapter, we have studied the theory of the hydrogen atom as a single-particle problem, an electron moving in a spherical potential. The hydrogen atom can be considered a two-particle system, an electron and a proton, with the Coulomb potential acting between them. In this chapter, we will show how to solve the Schrödinger equation for the wave function of the two-particle system. We will introduce the coordinates of the center of mass, which will allow us to split the Schrödinger equation into two independent equations, one for the center of mass motion and another for the relative motion of the electron and proton.

17.1 Center of Mass Motion

Consider a system composed of two particles of masses m_1 and m_2 moving in potential forces that depend on the positions of the particles and time, $V = V(\vec{r}_2 - \vec{r}_1, t)$, where $\vec{r}_2 = (x_2, y_2, z_2)$ and $\vec{r}_1 = (x_1, y_1, z_1)$ are position vectors of the particles.

Quantum Physics for Beginners
Zbigniew Ficek
Copyright © 2016 Pan Stanford Publishing Pte. Ltd.
ISBN 978-981-4669-38-2 (Hardcover), 978-981-4669-39-9 (eBook)
www.panstanford.com

The Schrödinger equation for the particles has the following form

$$-\frac{\hbar^2}{2m_1}\nabla_1^2\Psi - \frac{\hbar^2}{2m_2}\nabla_2^2\Psi + V(\vec{r}_2 - \vec{r}_1, t)\Psi = i\hbar\frac{\partial\Psi}{\partial t}, \quad (17.1)$$

where $\Psi = \Psi(\vec{r}_1, \vec{r}_2, t)$, and

$$\nabla_i^2 = \frac{\partial^2}{\partial x_i^2} + \frac{\partial^2}{\partial y_i^2} + \frac{\partial^2}{\partial z_i^2}, \quad i = 1, 2. \quad (17.2)$$

We may introduce coordinates of center of mass, which can allow us to reduce the wave equation for two particles into the wave equation for a single (global) particle moving in a potential field. The center of mass coordinates are defined as

$$X = \frac{m_1 x_1 + m_2 x_2}{m_1 + m_2}, \quad Y = \frac{m_1 y_1 + m_2 y_2}{m_1 + m_2}, \quad Z = \frac{m_1 z_1 + m_2 z_2}{m_1 + m_2}, \quad (17.3)$$

and a relative (mutual) distance between the particles

$$\vec{r} = \vec{r}_2 - \vec{r}_1, \quad (17.4)$$

where $x = x_2 - x_1$, $y = y_2 - y_1$, $z = z_2 - z_1$.

For simplicity of notation, denote by $M = m_1 + m_2$ the total mass of the particles, and $\Phi(\vec{R}, \vec{r}, t) = \Psi(\vec{r}_1, \vec{r}_2, t)$ in which $\vec{R} = \vec{R}(X, Y, Z)$ and $\vec{r} = \vec{r}(x, y, z)$.

We now transform the Schrödinger equation from the $\vec{r}_1$ and $\vec{r}_2$ coordinates to the center of mass coordinates, determined by $\vec{R}$ and $\vec{r}$. This requires to transform the Laplacians ∇_1^2 and ∇_2^2 to the new coordinates. Thus, using the chain rule, we get for the first-order derivative

$$\frac{\partial\Psi}{\partial x_1} = \frac{\partial\Phi}{\partial X}\frac{\partial X}{\partial x_1} + \frac{\partial\Phi}{\partial x}\frac{\partial x}{\partial x_1} = \frac{m_1}{M}\frac{\partial\Phi}{\partial X} - \frac{\partial\Phi}{\partial x}. \quad (17.5)$$

Then the second-order derivative is

$$\frac{\partial^2\Psi}{\partial x_1^2} = \frac{\partial}{\partial x_1}\frac{\partial\Psi}{\partial x_1} = \frac{\partial}{\partial x_1}\left(\frac{m_1}{M}\frac{\partial\Phi}{\partial X} - \frac{\partial\Phi}{\partial x}\right)$$

$$= \left(\frac{\partial X}{\partial x_1}\frac{\partial}{\partial X} + \frac{\partial x}{\partial x_1}\frac{\partial}{\partial x}\right)\left(\frac{m_1}{M}\frac{\partial\Phi}{\partial X} - \frac{\partial\Phi}{\partial x}\right)$$

$$= \left(\frac{m_1}{M}\right)^2\frac{\partial^2\Phi}{\partial X^2} - \frac{m_1}{M}\frac{\partial^2\Phi}{\partial X\partial x} - \frac{m_1}{M}\frac{\partial^2\Phi}{\partial x\partial X} + \frac{\partial^2\Phi}{\partial x^2}. \quad (17.6)$$

Thus

$$\frac{\partial^2\Psi}{\partial x_1^2} = \left(\frac{m_1}{M}\right)^2\frac{\partial^2\Phi}{\partial X^2} - \frac{2m_1}{M}\frac{\partial^2\Phi}{\partial X\partial x} + \frac{\partial^2\Phi}{\partial x^2}. \quad (17.7)$$

Similarly, for the y_1 and z_1 components

$$\frac{\partial^2 \Psi}{\partial y_1^2} = \left(\frac{m_1}{M}\right)^2 \frac{\partial^2 \Phi}{\partial Y^2} - \frac{2m_1}{M} \frac{\partial^2 \Phi}{\partial Y \partial y} + \frac{\partial^2 \Phi}{\partial y^2},$$

$$\frac{\partial^2 \Psi}{\partial z_1^2} = \left(\frac{m_1}{M}\right)^2 \frac{\partial^2 \Phi}{\partial Z^2} - \frac{2m_1}{M} \frac{\partial^2 \Phi}{\partial Z \partial z} + \frac{\partial^2 \Phi}{\partial z^2}. \qquad (17.8)$$

Thus, the Laplace operators ∇_1^2 and ∇_2^2 can be written as

$$\nabla_1^2 \Psi = \left(\frac{m_1}{M}\right)^2 \nabla_c^2 \Phi - \frac{2m_1}{M}\left(\frac{\partial^2 \Phi}{\partial X \partial x} + \frac{\partial^2 \Phi}{\partial Y \partial y} + \frac{\partial^2 \Phi}{\partial Z \partial z}\right) + \nabla^2 \Phi,$$

$$\nabla_2^2 \Psi = \left(\frac{m_2}{M}\right)^2 \nabla_c^2 \Phi + \frac{2m_2}{M}\left(\frac{\partial^2 \Phi}{\partial X \partial x} + \frac{\partial^2 \Phi}{\partial Y \partial y} + \frac{\partial^2 \Phi}{\partial Z \partial z}\right) + \nabla^2 \Phi,$$

$$(17.9)$$

where

$$\nabla_c^2 = \frac{\partial^2}{\partial X^2} + \frac{\partial^2}{\partial Y^2} + \frac{\partial^2}{\partial Z^2},$$

$$\nabla^2 = \frac{\partial^2}{\partial x^2} + \frac{\partial^2}{\partial y^2} + \frac{\partial^2}{\partial z^2}. \qquad (17.10)$$

Moreover,

$$\frac{1}{m_1}\nabla_1^2 \Psi + \frac{1}{m_2}\nabla_2^2 \Psi = \frac{1}{M}\nabla_c^2 \Phi + \frac{1}{\mu}\nabla^2 \Phi, \qquad (17.11)$$

where

$$\frac{1}{\mu} = \frac{1}{m_1} + \frac{1}{m_2} \qquad (17.12)$$

is the so-called reduced mass.

Thus, the Schrödinger equation can be written as

$$-\frac{\hbar^2}{2M}\nabla_c^2 \Phi - \frac{\hbar^2}{2\mu}\nabla^2 \Phi + V(\vec{r}, t)\Phi = i\hbar\frac{\partial \Phi}{\partial t}. \qquad (17.13)$$

Assume that the potential V is independent of time. Then the left-hand size of the Schrödinger equation depends solely on the spatial variables, whereas the right-hand size depends solely on time. In this case, both sizes are independent of each other but are equal to the same constant, say E_T. Each size can be solved separately.

The wave function can be written as a product of the spatial and time-dependent parts. Thus, it can be written as

$$\Phi(\vec{R}, \vec{r}, t) = \Phi(\vec{R}, \vec{r})\Phi(t) = \Phi(\vec{R}, \vec{r})e^{-\frac{i}{\hbar}E_T t}, \qquad (17.14)$$

where E_T is the total energy of the particles.

Hence, the Schrödinger equation reduces to

$$-\frac{\hbar^2}{2M}\nabla_c^2\Phi - \frac{\hbar^2}{2\mu}\nabla^2\Phi + [V(\vec{r}) - E_T]\Phi = 0. \qquad (17.15)$$

It is very similar in form to the Schrödinger equation for the hydrogen atom except it involves two Laplacians. Since the Laplacians act on separate variables, we may try to solve the Schrödinger equation using the method of separate variables, in which we write the wave function as

$$\Phi(\vec{R}, \vec{r}) = \Phi_c(\vec{R})\Phi_r(\vec{r}). \qquad (17.16)$$

In this case, the Schrodinger equation can be written as

$$-\frac{\hbar^2}{2M}\frac{1}{\Phi_c}\nabla_c^2\Phi_c = -\left[-\frac{\hbar^2}{2\mu}\frac{1}{\Phi_r}\nabla^2\Phi_r + V(\vec{r}) - E_T\right]. \qquad (17.17)$$

The left-hand side of the equation depends solely on the center of mass variables, whereas the right-hand side depends solely on $\vec{r}$. Therefore, the sides are independent of each other and are equal to a constant, say E_c. Hence, we can write Eq. (17.17) as two independent equations, each dependent only on one variable

$$-\frac{\hbar^2}{2M}\frac{1}{\Phi_c}\nabla_c^2\Phi_c = E_c,$$

$$-\left[-\frac{\hbar^2}{2\mu}\frac{1}{\Phi_r}\nabla^2\Phi_r + V(\vec{r}) - E_T\right] = E_c. \qquad (17.18)$$

The equations can be rewritten in the form

$$\frac{\hbar^2}{2M}\nabla_c^2\Phi_c + E_c\Phi_c = 0, \qquad (17.19)$$

$$-\frac{\hbar^2}{2\mu}\nabla^2\Phi_r + [V(\vec{r}) - E]\Phi_r = 0, \qquad (17.20)$$

where $E = E_T - E_c$. Let us summarize our findings.

Equation (17.19) is the equation of motion of a free particle of mass M. Thus, E_c is the kinetic energy of the center of mass. If the center of mass is stationary, then $E_c = 0$.

Equation (17.20) is of the same form as the Schrödinger equation for the hydrogen atom, a single particle of mass μ moving in the potential $V(\vec{r})$. We have solved this equation in the chapter on the quantum theory of hydrogen atom.

Tutorial Problems

Problem 17.1 Suppose that a particle of mass m can rotate around a fixed point A, such that $r = $ constant, and $\theta = \pi/2 = $ constant.

(a) Show that the motion of the particle is quantized.
(b) Show that the only acceptable solutions to the wave function of the particle are those corresponding to the positive energies $(E > 0)$ of the particle.

Chapter 18

Time-Independent Perturbation Theory

In many situations in physics, the Hamiltonian $\hat{H}$ of a given system is so complicated that the solution to the stationary Schrödinger equation is practically impossible or very difficult. Therefore, some approximation methods are required. In this chapter, we present the time-independent perturbation theory, the procedure of finding corrections to non-degenerate eigenvalues and eigenvectors to a part (called unperturbed part) of the Hamiltonian of a given system.

The perturbation theory is appropriate when the Hamiltonian can be split into two parts

$$\hat{H} = \hat{H}_0 + \hat{V}, \tag{18.1}$$

such that we can solve the eigenvalue equation for $\hat{H}_0$, i.e., we can find eigenvalues $E_n^{(0)}$ and eigenvectors $|\phi_n^{(0)}\rangle$ of the Hamiltonian $\hat{H}_0$, and we can treat the part $\hat{V}$ as a small perturber to $\hat{H}_0$.

Thus, the problem of solving the eigenvalue equation

$$\hat{H}|\phi\rangle = \left(\hat{H}_0 + \hat{V}\right)|\phi\rangle = E|\phi\rangle \tag{18.2}$$

reduces to find E and $|\phi\rangle$ when we know the eigenvalues $E_n^{(0)}$ and the eigenvectors $|\phi_n^{(0)}\rangle$ of $\hat{H}_0$.

Since $\hat{V}$ appears as a small perturber to $\hat{H}_0$, we will try to find E and $|\phi\rangle$ in the form of a series

$$|\phi\rangle = |\phi_n^{(0)}\rangle + |\phi_n^{(1)}\rangle + \dots,$$
$$E = E_n^{(0)} + E_n^{(1)} + \dots, \tag{18.3}$$

Quantum Physics for Beginners
Zbigniew Ficek
Copyright © 2016 Pan Stanford Publishing Pte. Ltd.
ISBN 978-981-4669-38-2 (Hardcover), 978-981-4669-39-9 (eBook)
www.panstanford.com

where $|\phi_n^{(1)}\rangle$ is the first-order correction to the unperturbed eigenstate $|\phi_n^{(0)}\rangle$, and $E_n^{(1)}$ is the first-order correction to the unperturbed eigenvalue $E_n^{(0)}$. The subscript n indicates that the Hamiltonian $\hat{H}_0$ can have more than one eigenvalue and eigenvector.

18.1 First-Order Corrections to Eigenvalues

Substituting the series expansion (18.3) into the eigenvalue equation (18.2), we obtain

$$\left(\hat{H}_0 + \hat{V}\right)\left(|\phi_n^{(0)}\rangle + |\phi_n^{(1)}\rangle\right) = \left(E_n^{(0)} + E_n^{(1)}\right)\left(|\phi_n^{(0)}\rangle + |\phi_n^{(1)}\rangle\right). \quad (18.4)$$

Expanding both sides of Eq. (18.4) and equating terms of the same order in $\hat{V}$, we obtain

$$\hat{H}_0|\phi_n^{(0)}\rangle = E_n^{(0)}|\phi_n^{(0)}\rangle, \qquad \text{zeroth order in } \hat{V}, \quad (18.5)$$

$$\hat{H}_0|\phi_n^{(1)}\rangle + \hat{V}|\phi_n^{(0)}\rangle = E_n^{(0)}|\phi_n^{(1)}\rangle + E_n^{(1)}|\phi_n^{(0)}\rangle, \quad \text{first order in } \hat{V}. \quad (18.6)$$

Equation (18.5) is the stationary Schrödinger equation whose solution is known. In order to solve Eq. (18.6), we write the equation in the form

$$\left(\hat{H}_0 - E_n^{(0)}\right)|\phi_n^{(1)}\rangle = E_n^{(1)}|\phi_n^{(0)}\rangle - \hat{V}|\phi_n^{(0)}\rangle. \quad (18.7)$$

Assume that the eigenvalues $E_n^{(0)}$ are non-degenerated, i.e., for a given $E_n^{(0)}$, there is only one eigenfunction $|\phi_n^{(0)}\rangle$.

Multiplying Eq. (18.7) from the left by $\langle\phi_n^{(0)}|$, we obtain

$$\langle\phi_n^{(0)}|\hat{H}_0|\phi_n^{(1)}\rangle - \langle\phi_n^{(0)}|E_n^{(0)}|\phi_n^{(1)}\rangle = E_n^{(1)}\langle\phi_n^{(0)}|\phi_n^{(0)}\rangle - \langle\phi_n^{(0)}|\hat{V}|\phi_n^{(0)}\rangle. \quad (18.8)$$

Since

$$\langle\phi_n^{(0)}|\hat{H}_0|\phi_n^{(1)}\rangle = \langle\phi_n^{(1)}|\hat{H}_0|\phi_n^{(0)}\rangle^* = E_n^{(0)}\langle\phi_n^{(0)}|\phi_n^{(1)}\rangle, \quad (18.9)$$

the left-hand side of Eq. (18.8) vanishes, giving

$$E_n^{(1)} = \langle\phi_n^{(0)}|\hat{V}|\phi_n^{(0)}\rangle = \langle n|\hat{V}|n\rangle. \quad (18.10)$$

Thus, the first-order correction to the eigenvalue $E_n^{(0)}$ is equal to the expectation value of $\hat{V}$ in the state $|\phi_n^{(0)}\rangle$.

18.2 First-Order Corrections to Eigenvectors

In order to find the first-order correction to the eigenstate $|\phi_n^{(0)}\rangle$, we expand $|\phi_n^{(1)}\rangle$ state in terms of $|\phi_m^{(0)}\rangle$, using the completeness relation as

$$|\phi_n^{(1)}\rangle = \sum_m |\phi_m^{(0)}\rangle\langle\phi_m^{(0)}|\phi_n^{(1)}\rangle = \sum_m c_{mn}|\phi_m^{(0)}\rangle, \qquad (18.11)$$

where $c_{mn} = \langle\phi_m^{(0)}|\phi_n^{(1)}\rangle$.

We find the coefficients c_{mn} from Eq. (18.7) by multiplying this equation from the left by $\langle\phi_m^{(0)}|$ $(m \neq n)$ and find

$$\langle\phi_m^{(0)}|\hat{H}_0|\phi_n^{(1)}\rangle - E_n^{(0)}\langle\phi_m^{(0)}|\phi_n^{(1)}\rangle = E_n^{(1)}\langle\phi_m^{(0)}|\phi_n^{(0)}\rangle - \langle\phi_m^{(0)}|\hat{V}|\phi_n^{(0)}\rangle.$$
$$(18.12)$$

Since

$$\langle\phi_m^{(0)}|\phi_n^{(0)}\rangle = 0$$

and

$$\langle\phi_m^{(0)}|\hat{H}_0|\phi_n^{(1)}\rangle = E_m^{(0)}\langle\phi_m^{(0)}|\phi_n^{(1)}\rangle, \qquad (18.13)$$

we get

$$c_{mn} = \langle\phi_m^{(0)}|\phi_n^{(1)}\rangle = \frac{\langle\phi_m^{(0)}|\hat{V}|\phi_n^{(0)}\rangle}{E_n^{(0)} - E_m^{(0)}}. \qquad (18.14)$$

Hence

$$|\phi_n^{(1)}\rangle = \sum_{m \neq n} \frac{\langle\phi_m^{(0)}|\hat{V}|\phi_n^{(0)}\rangle}{E_n^{(0)} - E_m^{(0)}} |\phi_m^{(0)}\rangle. \qquad (18.15)$$

Since we know $E_n^{(0)}$ and $|\phi_n^{(0)}\rangle$, we can find $E_n^{(1)}$ from Eq. (18.10) and $|\phi_n^{(1)}\rangle$ from Eq. (18.15).

The perturbation theory can be applied to analyze the quantum properties of particles trapped in two closely coupled potential wells. By closely coupled wells, it is meant that these two wells are separated by a barrier, as illustrated in Fig. 18.1. This is a typical situation in the studies of quantum dynamics of Bose–Einstein condensates. The perturbation theory is particularly useful here, as the following example will demonstrate.

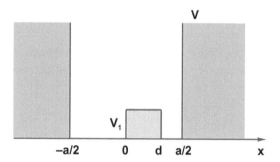

Figure 18.1 Infinite potential well with a small potential (perturber) barrier V_1.

Worked Example

Consider a particle in an infinite one-dimensional potential well, as shown in Fig. 18.1. Assume that inside the infinite well there is a small potential barrier of high V_1 and thickness d. Treating the barrier V_1 as a small perturber, find the eigenvalues and eigenstates of the particle valid to the first order in V_1.

Solution

We have learned in Section 8.1 that the eigenstates of the particle in the infinite well, without V_1, are

$$|\phi_n^{(0)}\rangle = \sqrt{\frac{2}{a}} \sin\left(n\frac{\pi x}{a}\right), \qquad (18.16)$$

with the corresponding eigenvalues

$$E_n^{(0)} = n^2 \frac{\pi^2 \hbar^2}{2m_p a^2}, \qquad (18.17)$$

where m_p is the mass of the particle.

Thus, the first-order correction to the eigenvalue $E_n^{(0)}$ is

$$E_n^{(1)} = \langle \phi_n^{(0)}|\hat{V}_1|\phi_n^{(0)}\rangle = \frac{2V_1}{a} \int_0^d dx \, \sin^2\left(\frac{n\pi x}{a}\right). \qquad (18.18)$$

To find the first-order correction to the eigenstate $|\phi_n^{(0)}\rangle$, we have to calculate the matrix element

$$V_{mn} = \langle \phi_m^{(0)} | \hat{V}_1 | \phi_n^{(0)} \rangle = \frac{2V_1}{a} \int_0^d dx \, \sin\left(\frac{m\pi x}{a}\right) \sin\left(\frac{n\pi x}{a}\right),$$

(18.19)

where $m \neq n$.

Performing the integrations in Eqs. (18.18) and (18.19), we get

$$E_n^{(1)} = \frac{V_1}{a}\left[d - \frac{1}{2a}\sin(2d\alpha)\right],$$

(18.20)

$$V_{mn} = \frac{V_1}{a}\left\{\frac{1}{\alpha - \beta}\sin[(\alpha - \beta)d] - \frac{1}{\alpha + \beta}\sin[(\alpha + \beta)d]\right\},$$

(18.21)

where $\alpha = n\pi/a$ and $\beta = m\pi/a$.

Hence, the first-order correction to the eigenstate $\phi_n^{(0)}$ is

$$|\phi_n^{(1)}\rangle = \frac{2m_p a^2}{\pi^2 \hbar^2} \sum_{m \neq n} \frac{V_{mn}}{n^2 - m^2} |\phi_m^{(0)}\rangle.$$

(18.22)

Tutorial Problems

Problem 18.1 In an orthonormal basis, a linear operator $\hat{A}$ is represented by a matrix

$$\hat{A} = \begin{pmatrix} 2\lambda & 1+\lambda \\ 1+\lambda & \lambda \end{pmatrix},$$

where λ is a small real parameter ($\lambda \ll 1$).

The operator $\hat{A}$ can be written as a sum of two operators $\hat{A} = \hat{A}_0 + \lambda\hat{V}$, where

$$\hat{A}_0 = \begin{pmatrix} 0 & 1 \\ 1 & 0 \end{pmatrix}, \quad \hat{V} = \begin{pmatrix} 2 & 1 \\ 1 & 1 \end{pmatrix}.$$

Using the first-order perturbation theory, find the eigenvalues and eigenvectors of $\hat{A}$ in terms of the eigenvalues and eigenvectors of $\hat{A}_0$.

Notice that $\hat{A}_0$ is the $\hat{\sigma}_x$ operator defined in Chapter 13.

Chapter 19

Time-Dependent Perturbation Theory

In this chapter, we continue our study of perturbation methods for the solution to the Schrödinger equation and turn our attention to the case where the perturber $\hat{V}$ depends explicitly on time. Just as in the case of the time-independent perturbation theory, we find the corrections to the eigenvalues and eigenvectors of a system described by a time-dependent Hamiltonian.

19.1 Iterative Method

In the time-independent perturbation theory, we have found corrections to the stationary (time-independent) eigenvalues and eigenstates of the Hamiltonian $\hat{H}$ by solving the stationary part of the Schrödinger equation

$$\hat{H}|\phi\rangle = E|\phi\rangle. \tag{19.1}$$

In the non-stationary (time-dependent) case, in which the Hamiltonian of the system depends on time, we find the time evolution of the state vector $\Psi(\vec{r}, t)$ by solving the time-dependent Schrödinger equation

$$i\hbar\frac{\partial}{\partial t}|\Psi(t)\rangle = \hat{H}(t)|\Psi(t)\rangle, \tag{19.2}$$

Quantum Physics for Beginners
Zbigniew Ficek
Copyright © 2016 Pan Stanford Publishing Pte. Ltd.
ISBN 978-981-4669-38-2 (Hardcover), 978-981-4669-39-9 (eBook)
www.panstanford.com

where for simplicity of the notation $|\Psi(t)\rangle \equiv |\Psi(\vec{r}, t)\rangle$ and $\hat{H}(t) \equiv \hat{H}(\vec{r}, t)$.

Direct integration of Eq. (19.2) gives

$$|\Psi(t)\rangle = |\Psi(0)\rangle + \frac{1}{i\hbar} \int_0^t dt' \, \hat{H}(t')|\Psi(t')\rangle. \qquad (19.3)$$

Since the right-hand side of Eq. (19.3) depends on $|\Psi(t')\rangle$, this is not the final solution to $|\Psi(t)\rangle$. The final solution to $|\Psi(t)\rangle$ should be, for example, in terms of the initial state $|\Psi(0)\rangle$.

To get the solution in this form, we can use the iteration method. In this approach, we substitute the solution given by Eq. (19.3) back into the right-hand side of the Schrödinger equation (19.2) and obtain

$$i\hbar \frac{\partial}{\partial t}|\Psi(t)\rangle = \hat{H}(t)|\Psi(0)\rangle + \frac{1}{i\hbar}\hat{H}(t) \int_0^t dt' \, \hat{H}(t')|\Psi(t')\rangle. \qquad (19.4)$$

Integrating the above equation, we get

$$|\Psi(t)\rangle = |\Psi(0)\rangle + \frac{1}{i\hbar} \int_0^t dt' \, \hat{H}(t')|\Psi(0)\rangle$$
$$+ \left(\frac{1}{i\hbar}\right)^2 \int_0^t dt' \int_0^{t'} dt'' \, \hat{H}(t')\hat{H}(t'')|\Psi(t'')\rangle. \qquad (19.5)$$

Proceeding further in this way, we obtain

$$|\Psi(t)\rangle = |\Psi(0)\rangle + \frac{1}{i\hbar} \int_0^t dt' \, \hat{H}(t')|\Psi(0)\rangle$$
$$+ \left(\frac{1}{i\hbar}\right)^2 \int_0^t dt' \int_0^{t'} dt'' \, \hat{H}(t')\hat{H}(t'')|\Psi(0)\rangle$$
$$\vdots$$
$$+ \left(\frac{1}{i\hbar}\right)^n \int_0^t dt' \ldots \int_0^{t_{n-1}} dt_n \, \hat{H}(t') \ldots \hat{H}(t_n)|\Psi(t_n)\rangle. \qquad (19.6)$$

Note the time ordering in the integration, $t \geq t' \geq t'' \geq \ldots t_n \geq 0$.

If the time t_n is short or the state $|\Psi(t_n)\rangle$ does not change much under the action of $\hat{H}(t') \ldots \hat{H}(t_n)$, we can approximate $|\Psi(t_n)\rangle$ by $|\Psi(0)\rangle$. Then, the iterative solution (19.6) can be accepted as the final solution to the wave function of the system.

The iterative solution (19.6) involves the total Hamiltonian of the system $\hat{H}(t)$. It may result in complicated expressions to evaluate.

A simpler way to proceed, which is a common practice in quantum physics, is to write the total Hamiltonian as the sum of two terms

$$\hat{H}(t) = \hat{H}_0 + \hat{V}(t), \tag{19.7}$$

where $\hat{H}_0$ is the time-independent (stationary) part of the Hamiltonian, and $\hat{V}(t)$ is the part containing all time-dependent terms, usually called the time-dependent interaction Hamiltonian or time-dependent perturber.

If $\hat{V}(0) = 0$, the state of the system at $t = 0$ is determined by the Hamiltonian $\hat{H}_0$, which satisfies the time-independent Schrödinger equation

$$\hat{H}_0|\Psi(0)\rangle = E_0|\Psi(0)\rangle. \tag{19.8}$$

In other words, the initial state of the system is given by the stationary eigenstates of $\hat{H}_0$.

The splitting of the total Hamiltonian into stationary and time-dependent parts allows to work in the interaction picture and find

$$|\Psi_I(t)\rangle = |\Psi(0)\rangle + \frac{1}{i\hbar}\int_0^t dt' \hat{V}(t')|\Psi(0)\rangle$$

$$+ \left(\frac{1}{i\hbar}\right)^2 \int_0^t dt' \int_0^{t'} dt'' \hat{V}(t')\hat{V}(t'')|\Psi(0)\rangle$$

$$\vdots$$

$$+ \left(\frac{1}{i\hbar}\right)^n \int_0^t dt' \ldots \int_0^{t_{n-1}} dt_n \hat{V}(t')\ldots\hat{V}(t_n)|\Psi(0)\rangle, \tag{19.9}$$

where we have used the fact that $|\Psi_I(0)\rangle = |\Psi(0)\rangle$.

19.2 Solution in Terms of Probability Amplitudes

The iterative solution (19.9) for the state vector of a system is in the operator form. We would prefer the solution in a number form rather than in the operator form. We can find the solution by finding first the unperturbed state of the system, $|\Psi_0(t)\rangle$, which is the solution to the Schrödinger equation

$$i\hbar\frac{\partial}{\partial t}|\Psi_0(t)\rangle = \hat{H}_0(t)|\Psi_0(t)\rangle. \tag{19.10}$$

The solution to Eq. (19.10) is of a simple form

$$|\Psi_0(t)\rangle = e^{-\frac{i}{\hbar}\hat{H}_0 t}|\Psi(0)\rangle. \tag{19.11}$$

Assume that $|\phi_n\rangle$ are the eigenstates of $\hat{H}_0$ with eigenvalues E_n. Then, we can write

$$|\Psi_0(t)\rangle = \sum_n c_n e^{-\frac{i}{\hbar}E_n t}|\phi_n\rangle, \tag{19.12}$$

where we have expanded the state vector $|\Psi(0)\rangle$ in terms of the eigenstates of $\hat{H}_0$:

$$|\Psi(0)\rangle = \sum_n c_n|\phi_n\rangle, \tag{19.13}$$

and next by the Taylor expansion of the exponent, we find

$$
\begin{aligned}
e^{-\frac{i}{\hbar}\hat{H}_0 t}|\phi_n\rangle &= \left[1 - \frac{i}{\hbar}\hat{H}_0 t + \frac{1}{2}\left(-\frac{i}{\hbar}\hat{H}_0 t\right)^2 + \ldots\right]|\phi_n\rangle \\
&= \left[1 - \frac{i}{\hbar}\hat{E}_n t + \frac{1}{2}\left(-\frac{i}{\hbar}\hat{E}_n t\right)^2 + \ldots\right]|\phi_n\rangle \\
&= e^{-\frac{i}{\hbar}E_n t}|\phi_n\rangle,
\end{aligned}
\tag{19.14}
$$

in which we have used the fact that $\hat{H}_0|\phi_n\rangle = E_n|\phi_n\rangle$.

If we write the Hamiltonian as the sum of two terms, Eq. (19.7), then the time-dependent Schrödinger equation can be written as

$$i\hbar\frac{\partial}{\partial t}|\Psi(t)\rangle = \hat{H}_0|\Psi(t)\rangle + \hat{V}(t)|\Psi(t)\rangle. \tag{19.15}$$

Since the right-hand side of Eq. (19.15) has two terms and we know the solution when only the first term is present, $\hat{H}_0|\Psi(t)\rangle$, we can solve the Schrödinger equation in the manner one solves an inhomogeneous differential equation. Namely, we may treat $\hat{H}_0|\Psi(t)\rangle$ as the homogeneous part and $\hat{V}(t)|\Psi(t)\rangle$ as the inhomogeneous part of the differential equation. Then, we can solve the equation by making the coefficients c_n dependent on time and write the solution as

$$|\Psi(t)\rangle = \sum_n c_n(t)e^{-\frac{i}{\hbar}E_n t}|\phi_n\rangle, \tag{19.16}$$

where $c_n(t)$ are unknown functions of time. To find the explicit time dependence of $c_n(t)$, we substitute Eq. (19.16) into Eq. (19.15) and get

$$i\hbar \sum_n \left(\dot{c}_n(t) - \frac{i}{\hbar} E_n c_n(t) \right) e^{-\frac{i}{\hbar} E_n t} |\phi_n\rangle$$
$$= \sum_n c_n(t) e^{-\frac{i}{\hbar} E_n t} E_n |\phi_n\rangle + \sum_n c_n(t) e^{-\frac{i}{\hbar} E_n t} \hat{V}(t) |\phi_n\rangle, \quad (19.17)$$

which simplifies to

$$i\hbar \sum_n \dot{c}_n(t) e^{-\frac{i}{\hbar} E_n t} |\phi_n\rangle = \sum_n c_n(t) e^{-\frac{i}{\hbar} E_n t} \hat{V}(t) |\phi_n\rangle. \quad (19.18)$$

Multiplying from the left by $\langle \phi_m |$, we obtain

$$i\hbar \dot{c}_m(t) e^{-\frac{i}{\hbar} E_m t} = \sum_n c_n(t) e^{-\frac{i}{\hbar} E_n t} V_{mn}(t), \quad (19.19)$$

where

$$V_{mn}(t) = \langle \phi_m | \hat{V}(t) | \phi_n \rangle, \quad (19.20)$$

and we have used the fact that $|\phi_n\rangle$ are orthonormal, $\langle \phi_m | \phi_n \rangle = \delta_{mn}$.

We see from Eq. (19.18) that the coefficients $c_n(t)$ satisfy a set of n ordinary differential equations

$$\dot{c}_m(t) = -\frac{i}{\hbar} \sum_n V_{mn}(t) c_n(t) e^{i\omega_{mn} t}, \quad (19.21)$$

where $\omega_{mn} = (E_m - E_n)/\hbar$.

In general, the set of the differential equations (19.21) can be solved exactly when $n \leq 4$. For $n > 4$ approximate methods are required. In the later case, we may find a solution by expanding $c_n(t)$ in powers of $V_{mn}(t)$:

$$c_n(t) = c_n^{(0)}(t) + c_n^{(1)}(t) + c_n^{(2)}(t) + \dots \quad (19.22)$$

Substituting Eq. (19.22) into Eq. (19.21), we get

$$\dot{c}_m^{(0)}(t) + \dot{c}_m^{(1)}(t) + \dot{c}_m^{(2)}(t) + \dots$$
$$= -\frac{i}{\hbar} \sum_n V_{mn}(t) e^{i\omega_{mn} t} \left(c_n^{(0)}(t) + c_n^{(1)}(t) + \dots \right). \quad (19.23)$$

Comparing the coefficients at the same powers of $V_{mn}(t)$, we obtain

$$\dot{c}_m^{(0)}(t) = 0,$$

$$\dot{c}_m^{(1)}(t) = -\frac{i}{\hbar} \sum_n V_{mn}(t) e^{i\omega_{mn}t} c_n^{(0)}(t),$$

$$\dot{c}_m^{(2)}(t) = -\frac{i}{\hbar} \sum_n V_{mn}(t) e^{i\omega_{mn}t} c_n^{(1)}(t),$$

$$\vdots$$

$$\dot{c}_m^{(p)}(t) = -\frac{i}{\hbar} \sum_n V_{mn}(t) e^{i\omega_{mn}t} c_n^{(p-1)}(t). \tag{19.24}$$

We see that if we know the zeroth-order coefficient $c_n^{(0)}(t)$, then we can find all the remaining coefficients by a successive substitution of the solution to the higher-order coefficient. Let us illustrate the procedure of the solution to Eq. (19.24) up to the second order, $c_n^{(2)}(t)$.

Integrating the differential equation for $c_m^{(0)}(t)$, we get

$$c_m^{(0)}(t) = c_m^{(0)}(0), \tag{19.25}$$

which is a constant independent of time. Assume that initially the system was in one of the eigenstates of $\hat{H}_0$, say $|\phi_k\rangle$. Then

$$c_m^{(0)}(0) = \langle \phi_m | \phi_k \rangle = \delta_{mk}. \tag{19.26}$$

Substituting the solution (19.26) into the differential equation for $\dot{c}_m^{(1)}(t)$, we get

$$\dot{c}_m^{(1)}(t) = -\frac{i}{\hbar} V_{mk}(t) e^{i\omega_{mk}t}. \tag{19.27}$$

An integration gives

$$c_m^{(1)}(t) = -\frac{i}{\hbar} \int_0^t dt' V_{mk}(t') e^{i\omega_{mk}t'}. \tag{19.28}$$

Note that k is the initial and m is the final state of the system. If $V_{mk}(t)$ is independent of time, then

$$c_m^{(1)}(t) = -\frac{i}{\hbar} V_{mk} \frac{1}{i\omega_{mk}} \left(e^{i\omega_{mk}t} - 1 \right)$$

$$= -\frac{V_{mk}}{E_m - E_k} \left(e^{i\omega_{mk}t} - 1 \right). \tag{19.29}$$

Thus, the first-order correction to the probability amplitude is proportional to V_{mk}.

To find the second-order correction, we substitute the solution to $c_m^{(1)}(t)$, Eq. (19.29), into the differential equation for $c_m^{(2)}(t)$ in Eq. (19.24) and obtain

$$\dot{c}_m^{(2)}(t) = \frac{i}{\hbar} \sum_n \frac{V_{mn} V_{nk}}{E_n - E_k} \left(e^{i\omega_{nk}t} - 1 \right) e^{i\omega_{mn}t}$$

$$= \frac{i}{\hbar} \sum_n \frac{V_{mn} V_{nk}}{E_n - E_k} \left(e^{i\omega_{mk}t} - e^{i\omega_{mn}t} \right). \qquad (19.30)$$

Note, since m is the final state and k is the initial state, the sum over n is the summation over all intermediate states.

Integrating Eq. (19.30), we get

$$c_m^{(2)}(t) = \sum_n \frac{V_{mn} V_{nk}}{E_n - E_k} \left[\frac{e^{i\omega_{mk}t} - 1}{E_m - E_k} - \frac{e^{i\omega_{mn}t} - 1}{E_m - E_n} \right]. \qquad (19.31)$$

Worked Example

Consider a harmonic oscillator with a time-independent perturber

$$\hat{V} = \alpha \left(\hat{a} + \hat{a}^\dagger \right), \qquad (19.32)$$

where α is a small real constant. Find the first- and second-order corrections to the probability amplitude of the transition $|0\rangle \rightarrow |2\rangle$.

Solution

In order to find the first-order correction to the probability amplitude, we have to calculate the matrix element V_{20}. From Eq. (19.32) and using the properties of the annihilation and creation operators, $\hat{a}|0\rangle = 0$, $\hat{a}^\dagger|0\rangle = |1\rangle$, we find

$$V_{20} = \langle 2|\hat{V}|0\rangle = \alpha\langle 2|(\hat{a} + \hat{a}^\dagger)|0\rangle = \alpha\langle 2|1\rangle = 0. \qquad (19.33)$$

Thus, the first-order correction to the transition amplitude is zero.

For the second-order correction, we need matrix elements V_{21} and V_{10}, which are found as

$$V_{21} = \langle 2|\hat{V}|1\rangle = \alpha\langle 2|(\hat{a} + \hat{a}^\dagger)|1\rangle = \alpha\langle 2|2\rangle = \alpha,$$
$$V_{10} = \langle 1|\hat{V}|0\rangle = \alpha\langle 1|(\hat{a} + \hat{a}^\dagger)|0\rangle = \alpha\langle 1|1\rangle = \alpha. \quad (19.34)$$

Hence

$$c_2^{(2)}(t) = \frac{\alpha^2}{E_1 - E_0}\left[\frac{e^{i\omega_{20}t} - 1}{E_2 - E_0} - \frac{e^{i\omega_{21}t} - 1}{E_2 - E_1}\right]. \quad (19.35)$$

19.3 Transition Probability

To the first order, the transition amplitude (probability amplitude) is given by

$$c_m^{(1)}(t) \equiv c_{k\to m}^{(1)}(t) = -\frac{V_{mk}}{E_m - E_n}\left(e^{i\omega_{mk}t} - 1\right). \quad (19.36)$$

Then the transition probability is

$$P_{k\to m}(t) = |c_{k\to m}^{(1)}(t)|^2 = \left|\frac{V_{mk}}{E_m - E_n}\left(e^{i\omega_{mk}t} - 1\right)\right|^2. \quad (19.37)$$

Since

$$\left|e^{i\omega_{mk}t} - 1\right|^2 = 4\sin^2\left(\frac{1}{2}\omega_{mk}t\right), \quad (19.38)$$

the transition probability simplifies to

$$P_{k\to m}(t) = 4\left|\frac{V_{mk}}{E_m - E_n}\right|^2 \sin^2\left(\frac{1}{2}\omega_{mk}t\right). \quad (19.39)$$

Notice the reversibility of the transition probability

$$P_{k\to m}(t) = P_{m\to k}(t). \quad (19.40)$$

Thus, for the same system and in the same time interval, the transitions $k \to m$ and $m \to k$ occur with the same probability. This property is known in the literature as the *principle of detailed balance*.

Let us look at the transition probability $P_{k\to m}(t)$ more closely. Since

$$E_m - E_k = \hbar\omega_{mk}, \quad (19.41)$$

the expression (19.39) for the transition probability reduces to

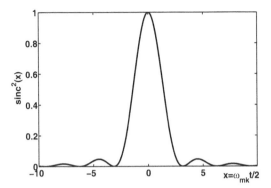

Figure 19.1 The variation of the function $\sin c^2 x$ with $x = \frac{1}{2}\omega_{mk}t$.

$$P_{k \to m}(t) = \frac{|V_{mk}|^2}{\hbar^2} \frac{\sin^2 \left(\frac{1}{2}\omega_{mk}t\right)}{\left(\frac{1}{2}\omega_{mk}\right)^2} = \frac{|V_{mk}|^2 t^2}{\hbar^2} \frac{\sin^2 \left(\frac{1}{2}\omega_{mk}t\right)}{\left(\frac{1}{2}\omega_{mk}t\right)^2}. \quad (19.42)$$

We see that the probability exhibits a $\sin c^2 x$ behavior, where $\sin cx = \sin x/x$, and $x = \omega_{mk}t/2$.

Figure 19.1 shows the variation of $\sin c^2 x$ function with x. We see that the function has a pronounced peak centered at $x = 0$ ($\omega_{mk} = 0$). The zeroth of the function are at $x = \pm n\pi$, ($n = 1, 2, \ldots$). The width of the peak is $2\pi/t$ showing that the probability depends inversely on the transition (observation) time. The width of the peak narrows with an increasing transition time. This is consistent with the time–energy uncertainty relation.

For short times ($x \ll 1$), $\sin c^2 x \approx 1$, and then

$$P_{k \to m}(t) \approx \frac{|V_{mk}|^2}{\hbar^2} t^2. \quad (19.43)$$

Notice the proportionality of the transition probability to t^2.

19.3.1 *Fermi Golden Rule*

The result (19.43) that the probability varies quadratically with time is not consistent with experimental observations. The experiments show that the probability is proportional to t rather than to t^2.

Where this inconsistency comes from?

In the derivation of $P_{k \to m}(t)$, we have taken into account only two energy levels, k and m. However, electrons in atoms can be in many energy levels, in fact, in an infinite number of levels. For a short transition time, the uncertainty of energy to which the transition occurs is very large. Thus, we cannot be sure to which atomic level the electron can be transferred over a short time t. In this case, we have to assume that after an excitation, the atom can be found in one of the large number of excited levels. Hence, the probability of an excitation of the atom to one of its energy levels is the sum of the probabilities of transitions from k to all the excited levels

$$P_{k \to m}(t) = \sum_m \left| c_{k \to m}^{(1)}(t) \right|^2. \tag{19.44}$$

If the number of energy levels is large, we may approximate them by a continuum of levels and replace the sum by an integral

$$P_{k \to m}(t) = \int_E^{E+dE} \left| c_{k \to m}^{(1)}(t) \right|^2 dm, \tag{19.45}$$

where dm is the number of energy levels in the energy width dE.

We can express the number of energy levels in the continuum by the density of energy levels in the continuum

$$dm = \rho(\omega_{mk}) dE = \hbar \rho(\omega_{mk}) d\omega_{mk}, \tag{19.46}$$

where $\rho(\omega_{mk})$ is the density of the energy levels. Hence

$$
\begin{aligned}
P_{k \to m}(t) &= \hbar \int_{-\infty}^{\infty} d\omega_{mk} \rho(\omega_{mk}) \left| c_{k \to m}^{(1)}(t) \right|^2 \\
&= \int_{-\infty}^{\infty} d\omega_{mk} \rho(\omega_{mk}) \frac{|V_{mk}|^2 t^2 \sin^2\left(\frac{1}{2}\omega_{mk}t\right)}{\hbar \left(\frac{1}{2}\omega_{mk}t\right)^2}.
\end{aligned} \tag{19.47}
$$

If V_{mk} and $\rho(\omega_{mk})$ vary slowly with ω_{mk}, we may consider them as constant functions over ω_{mk} and take them out of the integral. Then, we obtain

$$
\begin{aligned}
P_{k \to m}(t) &= \frac{|V_{mk}|^2 t^2}{\hbar} \rho(\omega_{mk}) \int_{-\infty}^{\infty} d\omega_{mk} \frac{\sin^2\left(\frac{1}{2}\omega_{mk}t\right)}{\left(\frac{1}{2}\omega_{mk}t\right)^2} \\
&= \frac{2|V_{mk}|^2 t}{\hbar} \rho(\omega_{mk}) \int_{-\infty}^{\infty} dx \frac{\sin^2 x}{x^2},
\end{aligned} \tag{19.48}
$$

where $x = \omega_{mk}t/2$. Since

$$\int_{-\infty}^{\infty} dx \frac{\sin^2 x}{x^2} = \pi, \tag{19.49}$$

we finally get the following expression for the transition probability:

$$P_{k \to m}(t) = \frac{2\pi |V_{mk}|^2}{\hbar} \rho(\omega_{mk}) t. \tag{19.50}$$

This is the famous *Fermi Golden Rule*, which shows that in an atom composed of a large number of energy states, the probability of a transition is proportional to time. Thus, per unit time of the transition probability is constant and proportional to $|V_{mk}|^2$.

An interesting observation: *Shift of the energy levels*

Consider the probability amplitude of the $k \to m$ transition for $\hat{V}$ independent of time. To the first order that the electron makes, a transition from k to m is

$$c_{k \to m}(t) = c_{k \to m}^{(0)}(t) + c_{k \to m}^{(1)}(t) = \delta_{km} - \frac{V_{mk}}{E_m - E_k} \left(e^{i\omega_{mk}t} - 1 \right). \tag{19.51}$$

For short times ($t \ll 1$), the transition amplitude (19.51) reduces to

$$c_{k \to m}(t) = \delta_{km} - \frac{i}{\hbar} V_{mk} t. \tag{19.52}$$

The transition amplitude exhibits interesting properties. Namely, the probability amplitude that the system will remain in the state k is

$$c_{kk}(t) \equiv c_{k \to k}(t) = 1 - \frac{i}{\hbar} V_{kk} t. \tag{19.53}$$

The probability amplitude that the system will make a transition to a state m is

$$c_{km}(t) \equiv c_{k \to m}(t) = -\frac{i}{\hbar} V_{mk} t. \tag{19.54}$$

Note that the amplitudes $c_{kk}(t)$ and $c_{km}(t)$ depend on different matrix elements V_{kk} and V_{mk}, respectively. Thus, even if the system does not make transitions to other states ($V_{mk} = 0$), the probability that the system remains in the state k may change in time.

How to understand it?

In the derivation of $c_{km}(t)$, we have assumed that $\hat{V}$ is independent of time. Hence, the time-dependent perturbation theory should lead to results one could expect from the time-independent perturbation theory.

In the time-independent perturbation theory, V_{kk} contributes to the first-order correction to the energy of the state k. Thus, even if the system does not make transitions to other states, the perturber can still perturb (shift) energy of the initial state.

In order to show that there is a shift of the unperturbed energy E_k, we consider the exact equation of motion for $c_k(t)$:

$$\dot{c}_k(t) = -\frac{i}{\hbar} \sum_n V_{kn} c_n(t) e^{i\omega_{kn}t}. \tag{19.55}$$

We may extract from the sum over n the term $n = k$, so we can write

$$\dot{c}_k(t) = -\frac{i}{\hbar} V_{kk} c_k(t) - \frac{i}{\hbar} \sum_{n \neq k} V_{kn} c_n(t) e^{i\omega_{kn}t}. \tag{19.56}$$

If the system does not make transitions to $n \neq k$, i.e., $V_{kn} = 0$, and then

$$\dot{c}_k(t) = -\frac{i}{\hbar} V_{kk} c_k(t). \tag{19.57}$$

The solution to Eq. (19.57) is

$$c_k(t) = c_k(0) e^{-\frac{i}{\hbar} V_{kk} t}. \tag{19.58}$$

Substituting the solution (19.58) into the expression for the state vector, Eq. (19.16), we obtain

$$|\Psi(t)\rangle = \sum_n c_n(t) e^{-\frac{i}{\hbar} E_n t} |\phi_n\rangle = \sum_n c_n(0) e^{-\frac{i}{\hbar}(E_n + V_{nn})t} |\phi_n\rangle. \tag{19.59}$$

This clearly shows that the energy of the nth state of the system is shifted from its unperturbed value E_n by V_{nn}.

An useful comment: How do in practice we observe (detect) that an electron makes transitions between different energy states?

When the electron makes a transition from a lower energy state to a higher energy state, an energy is absorbed from the external perturber, and vice versa, when the electron makes a transition from a higher energy state to a lower energy state, an energy is emitted. In a laboratory, we measure the absorbed or emitted energy. Usually, the absorbed or emitted energy is in the form of a radiation. A detector measures the square of the amplitude of the field falling on it, $|E|^2$, whose average value $\langle |E|^2 \rangle$ is called intensity.

We can show, using the time-dependent perturbation theory, that the intensity of the absorbed or emitted radiation is proportional to the probability of the transition between two energy states of the absorbing or emitting system.

The probability that a system will be found at time t in a state $|m\rangle$ is given by

$$P_m(t) = |\langle m|\Psi(t)\rangle|^2, \tag{19.60}$$

where $|\Psi(t)\rangle$ is the state vector of the system. If the system interacts with an external perturber described by an operator $\hat{V}_I$, we know from the perturbation theory that to the first order in the perturbation, the state vector of the system evolves in time as

$$|\Psi(t)\rangle = |\Psi(0)\rangle + \frac{1}{i\hbar}\int_0^t dt'\, \hat{V}_I(t')|\Psi(0)\rangle. \tag{19.61}$$

Suppose that the external perturber is linearly coupled to the system, i.e., the interaction Hamiltonian is of the form

$$\hat{V}_I(t) = -i\hbar\left[\mathcal{E}(t)S_{mk} - \mathcal{E}^*(t)S_{km}\right], \tag{19.62}$$

where $\mathcal{E}(t)$ is the amplitude of the external perturber, which by acting on the system forces the electron to make a transition between the energy states k and m. The transitions are described by the transition (projection) operators $S_{mk} = |m\rangle\langle k|$ for the "up" $k \rightarrow m$ transition, and $S_{km} = |k\rangle\langle m|$ for the "down" transition $m \rightarrow k$.

Assume that the initial state of the system plus the detected field is a product state, i.e., the system and the detected field are initially independent of each other

$$|\Psi(0)\rangle = |\Psi_F\rangle|k\rangle, \tag{19.63}$$

where $|\Psi_F\rangle$ is the initial state of the detected field. Then, we find using Eqs. (19.61) and (19.62) that a projection of $|m\rangle$ on $|\Psi(t)\rangle$ results in a state

$$\langle m|\Psi(t)\rangle = -\mathcal{E}(t)|\Psi_F\rangle t. \tag{19.64}$$

Hence, the probability that under the interaction with the perturber the system makes a transition to the state $|m\rangle$ is

$$P_m(t) = |\langle m|\Psi(t)\rangle|^2 = \langle \Psi_F|\mathcal{E}^*(t)\mathcal{E}(t)|\Psi_F\rangle t^2$$
$$= \langle \Psi_F||\mathcal{E}(t)|^2|\Psi_F\rangle t^2 = I_F(t)t^2, \tag{19.65}$$

where $I_F(t) = \langle \Psi_F||\mathcal{E}(t)|^2|\Psi_F\rangle$ is the intensity of the detected field.

Clearly, the probability of the $k \rightarrow m$ transition in the system is proportional to the intensity of the absorbed field.

Tutorial Problems

Problem 19.1 Consider a two-level atom represented by the spin operators $\hat{\sigma}^{\pm}$, $\hat{\sigma}_z$, interacting with a one-dimensional harmonic oscillator, represented by the creation and annihilation operators $\hat{a}^{\dagger}$ and $\hat{a}$. The Hamiltonian of the system is given by

$$\hat{H} = \frac{1}{2}\hbar\omega_0\hat{\sigma}_z + \hbar\omega_0\left(\hat{a}^{\dagger}\hat{a} + \frac{1}{2}\right) - \frac{1}{2}i\hbar g\left(\hat{\sigma}^{+}\hat{a} - \hat{\sigma}^{-}\hat{a}^{\dagger}\right). \quad (19.66)$$

The Hamiltonian can be written as

$$\hat{H} = \hat{H}_0 + \hat{V}, \quad (19.67)$$

where

$$\hat{H}_0 = \frac{1}{2}\hbar\omega_0\hat{\sigma}_z + \hbar\omega_0\left(\hat{a}^{\dagger}\hat{a} + \frac{1}{2}\right),$$

$$\hat{V} = -\frac{1}{2}i\hbar g\left(\hat{\sigma}^{+}\hat{a} - \hat{\sigma}^{-}\hat{a}^{\dagger}\right). \quad (19.68)$$

The eigenstates of $\hat{H}_0$ are product states

$$|\phi_n\rangle = |n\rangle|1\rangle, \quad |\phi_{n-1}\rangle = |n-1\rangle|2\rangle, \quad (19.69)$$

where $|n\rangle$ is the photon number state of the harmonic oscillator and $|1\rangle$, $|2\rangle$ are the energy states of the atom.

(a) Write the state vector of the system in terms of the eigenstates of $\hat{H}_0$.

(b) Assume that initially at $t = 0$, the system was in the state $|\phi_n\rangle$. Find the probability, using the time-dependent perturbation theory, that after a time t, the system can be found in the state $|\phi_{n-1}\rangle$.

Chapter 20

Relativistic Schrödinger Equation

Thus far our principal task has been the development of the fundamentals of quantum physics for non-relativistic (stationary or slowly moving) particles. Now we proceed to introduce the concepts of the relativistic theory to basic problems of quantum physics, in particular, the extension of the Schrödinger equation to a relativistic form. We will see that this is not an easy task and faces a considerable complication. We will find a controversy of how to implement the concepts of relativity to derive a relativistic form of the Schrödinger equation. Starting from the fundamental laws of relativity, we first derive the Klein–Gordon equation and investigate if the equation can be considered a generalization of the non-relativistic Schrödinger equation to the case of the relativistic energy. We will find that the wave function, which is a solution to the Klein–Gordon equation, cannot be connected with the probability wave function. For this reason, we will derive the Dirac equation, which solves the problem faced by the Klein–Gordon equation. The Dirac equation also includes the spin.

Consider a free particle for which the non-relativistic Schrödinger equation

$$\hat{H}\Psi = E\Psi, \tag{20.1}$$

Quantum Physics for Beginners
Zbigniew Ficek
Copyright © 2016 Pan Stanford Publishing Pte. Ltd.
ISBN 978-981-4669-38-2 (Hardcover), 978-981-4669-39-9 (eBook)
www.panstanford.com

is obtained from the non-relativistic formula for the energy (Hamiltonian)

$$\hat{H} = \frac{p^2}{2m},$$ (20.2)

by using Jordan's rules

$$E \rightarrow -\frac{\hbar}{i}\frac{\partial}{\partial t}, \qquad \vec{p} = \frac{\hbar}{i}\nabla.$$ (20.3)

When applying Jordan's rules to the Hamiltonian (20.2), we get

$$\hat{H} = \frac{p^2}{2m} = -\frac{\hbar^2}{2m}\nabla^2,$$ (20.4)

and then the Schrodinger equation takes the form

$$i\hbar\frac{\partial}{\partial t}\Psi + \frac{\hbar^2}{2m}\nabla^2\Psi = 0.$$ (20.5)

20.1 Klein–Gordon Equation

Consider now the relativistic formula for the energy of a free particle

$$E^2 = c^2 p^2 + m^2 c^4,$$ (20.6)

that

$$E = \pm\sqrt{c^2 p^2 + m^2 c^4}.$$ (20.7)

The formula (20.7) is not convenient for the quantization because there is minus sign in the square root. This indicates that even before the quantization, the classical kinetic energy could be negative.

Therefore, we consider the square of the energy, as given by Eq. (20.6), which is positive and write it as

$$E^2 - c^2 p^2 - m^2 c^4 = 0.$$ (20.8)

We now apply Jordan's rules. Since the relativistic energy formula involves E^2 and p^2, taking the square in Eq. (20.3), we obtain

$$E^2 = -\hbar^2\frac{\partial^2}{\partial t^2}, \qquad p^2 = -\hbar^2\nabla^2.$$ (20.9)

Then Eq. (20.8) takes the form

$$-\hbar^2\frac{\partial^2}{\partial t^2} + c^2\hbar^2\nabla^2 - m^2 c^4 = 0,$$ (20.10)

which can be rewritten as

$$\frac{\partial^2}{\partial t^2} - c^2 \nabla^2 + \frac{m^2 c^4}{\hbar^2} = 0, \tag{20.11}$$

or

$$\frac{1}{c^2} \frac{\partial^2}{\partial t^2} - \nabla^2 + \frac{m^2 c^2}{\hbar^2} = 0. \tag{20.12}$$

Introducing the notation

$$\frac{1}{c^2} \frac{\partial^2}{\partial t^2} - \nabla^2 \equiv \Box, \tag{20.13}$$

which is a four-vector (t, x, y, z) operator, called the d'Alembertian operator, we obtain a wave equation

$$\left(\Box + \frac{m^2 c^2}{\hbar^2} \right) \Psi = 0. \tag{20.14}$$

This equation is called the Klein–Gordon wave equation.

20.2 Difficulties of the Klein–Gordon Equation

A question arises: Can we consider the Klein–Gordon equation as a relativistic form of the Schrödinger equation?

If yes, then $|\Psi|^2$ should correspond to the probability density and then the wave function Ψ could be interpreted as a probability wave function. Let us check if $|\Psi|^2$ could be interpreted as the probability density.

This would be the case if the Klein–Gordon equation could be transformed to the continuity equation

$$\frac{\partial}{\partial t} \rho + \operatorname{div} \vec{j} = 0. \tag{20.15}$$

Let us try to transform the Klein–Gordon equation into the continuity equation. From the Klein–Gordon equation, we get

$$\Psi^* \Psi - (\Psi)^* \Psi = 0, \tag{20.16}$$

which can be written as

$$\Psi^* \left(\frac{1}{c^2} \frac{\partial^2 \Psi}{\partial t^2} - \nabla^2 \Psi \right) - \left(\frac{1}{c^2} \frac{\partial^2 \Psi^*}{\partial t^2} - \nabla^2 \Psi^* \right) \Psi = 0. \tag{20.17}$$

Grouping the time-dependent terms and spatial terms, the above equation takes the form

$$\frac{1}{c^2}\left[\Psi^*\frac{\partial^2\Psi}{\partial t^2}-\Psi\frac{\partial^2\Psi^*}{\partial t^2}\right]+\left[(\nabla^2\Psi^*)\,\Psi-\Psi^*\nabla^2\Psi\right]=0. \quad (20.18)$$

Since

$$\frac{\partial}{\partial t}\left(\Psi^*\frac{\partial\Psi}{\partial t}\right)=\frac{\partial\Psi^*}{\partial t}\frac{\partial\Psi}{\partial t}+\Psi^*\frac{\partial^2\Psi}{\partial t^2}, \quad (20.19)$$

and

$$\frac{\partial}{\partial t}\left(\Psi\frac{\partial\Psi^*}{\partial t}\right)=\frac{\partial\Psi}{\partial t}\frac{\partial\Psi^*}{\partial t}+\Psi\frac{\partial^2\Psi^*}{\partial t^2}, \quad (20.20)$$

we have

$$\Psi^*\frac{\partial^2\Psi}{\partial t^2}-\Psi\frac{\partial^2\Psi^*}{\partial t^2}=\frac{\partial}{\partial t}\left(\Psi^*\frac{\partial\Psi}{\partial t}\right)-\frac{\partial\Psi^*}{\partial t}\frac{\partial\Psi}{\partial t}-\frac{\partial}{\partial t}\left(\Psi\frac{\partial\Psi^*}{\partial t}\right)$$
$$+\frac{\partial\Psi}{\partial t}\frac{\partial\Psi^*}{\partial t}. \quad (20.21)$$

Hence

$$\Psi^*\frac{\partial^2\Psi}{\partial t^2}-\Psi\frac{\partial^2\Psi^*}{\partial t^2}=\frac{\partial}{\partial t}\left(\Psi^*\frac{\partial\Psi}{\partial t}-\Psi\frac{\partial\Psi^*}{\partial t}\right). \quad (20.22)$$

Similarly

$$(\nabla^2\Psi^*)\,\Psi-\Psi^*\nabla^2\Psi=-\nabla\cdot(\Psi^*\nabla\Psi-\Psi\nabla\Psi^*). \quad (20.23)$$

Thus, we obtain

$$\frac{\partial}{\partial t}\left[\frac{1}{c^2}\left(\Psi^*\frac{\partial\Psi}{\partial t}-\Psi\frac{\partial\Psi^*}{\partial t}\right)\right]-\nabla\cdot(\Psi^*\nabla\Psi-\Psi\nabla\Psi^*)=0, \quad (20.24)$$

or

$$\frac{\partial}{\partial t}\left[\frac{i\hbar}{2mc^2}\left(\Psi^*\frac{\partial\Psi}{\partial t}-\Psi\frac{\partial\Psi^*}{\partial t}\right)\right]+\nabla\cdot\left[\frac{\hbar}{2im}(\Psi^*\nabla\Psi-\Psi\nabla\Psi^*)\right]=0. \quad (20.25)$$

This equation has a form of the continuity equation, so that

$$\rho=\frac{i\hbar}{2mc^2}\left(\Psi^*\frac{\partial\Psi}{\partial t}-\Psi\frac{\partial\Psi^*}{\partial t}\right),$$

$$\vec{J}=\frac{\hbar}{2im}(\Psi^*\nabla\Psi-\Psi\nabla\Psi^*). \quad (20.26)$$

Note that ρ does not have to be positive. Therefore, ρ cannot be interpreted as the probability density.

Therefore, we cannot connect the solution to the Klein–Gordon equation with probability. The source of this problem is in the fact that the Klein–Gordon equation is second order in t, whereas the Schrödinger equation is first order in t.

Note that the Klein–Gordon equation for a free particle has only *one* solution for Ψ, i.e., gives only a single value of Ψ. For this reason, the Klein–Gordon equation cannot be an equation for a particle that has a nonzero spin, e.g., electron, as it cannot take either $+$ or $-$ value. Therefore, the Klein–Gordon equation can describe particles without spin.

20.3 Dirac Equation

We now turn to obtain a relativistic wave equation, which could be applicable for an arbitrary particle that may or may not possess a spin and the wave function could be interpreted as a probability wave function.

Consider a wave function of a particle whose spin could have different values

$$\Psi = \begin{pmatrix} \Psi_1 \\ \Psi_2 \\ . \\ . \\ \Psi_N \end{pmatrix}, \tag{20.27}$$

i.e., particle that can exist in N different states.

Define the probability density

$$\rho = \sum_{s=1}^{N} |\Psi_s|^2, \tag{20.28}$$

and consider the Schrödinger equation

$$i\hbar \frac{\partial}{\partial t} \Psi = H \Psi. \tag{20.29}$$

In quantum mechanics, we require for the Hamiltonian to be always linear and Hermitian. We may choose the Hamiltonian of the form

$$H = c\vec{\alpha} \cdot \vec{p} + \beta mc^2, \tag{20.30}$$

where $(\vec{\alpha}, \beta)$ are four operators acting on the components of the wave function. The explicit form of the operators has to be determined such that the Hamiltonian H satisfies the Schrödinger equation.

If we now apply Jordan's rules to the Hamiltonian (20.30), we get for the Schrödinger equation $H\Psi = E\Psi$:

$$\left(-\frac{\hbar}{i}\frac{\partial}{\partial t} - c\vec{\alpha} \cdot \left(\frac{\hbar}{i}\right)\nabla - \beta mc^2\right)\Psi = 0, \qquad (20.31)$$

which can be written as

$$\left(\frac{\hbar}{i}\frac{\partial}{\partial t} + \frac{\hbar c}{i}\vec{\alpha} \cdot \nabla + \beta mc^2\right)\Psi = 0, \qquad (20.32)$$

or equivalently

$$\left(E - c\vec{\alpha} \cdot \vec{p} - \beta mc^2\right)\Psi = 0. \qquad (20.33)$$

Since in the relativistic theory

$$E^2 = c^2 p^2 + m^2 c^4, \qquad (20.34)$$

we require that the solution to either Eq. (20.32) or Eq. (20.33) satisfy the relation for E, Eq. (20.34).

How to achieve it?

Let us act on Eq. (20.33) with the operator

$$E + c\vec{\alpha} \cdot \vec{p} + \beta mc^2, \qquad (20.35)$$

and get

$$\left(E - c\vec{\alpha} \cdot \vec{p} - \beta mc^2\right)\left(E + c\vec{\alpha} \cdot \vec{p} + \beta mc^2\right)\Psi = 0, \qquad (20.36)$$

which after performing the multiplication takes the form

$$\left\{E^2 - c^2(\vec{\alpha} \cdot \vec{p})^2 - mc^3[(\vec{\alpha} \cdot \vec{p})\beta + \beta(\vec{\alpha} \cdot \vec{p})] - \beta^2 m^2 c^4\right\}\Psi = 0. \qquad (20.37)$$

However,

$$\vec{\alpha} \cdot \vec{p} = \alpha_x p_x + \alpha_y p_y + \alpha_z p_z, \qquad (20.38)$$

so that

$$\begin{aligned}
(\vec{\alpha} \cdot \vec{p})^2 &= (\alpha_x p_x + \alpha_y p_y + \alpha_z p_z)(\alpha_x p_x + \alpha_y p_y + \alpha_z p_z) \\
&= \alpha_x^2 p_x^2 + \alpha_y^2 p_y^2 + \alpha_z^2 p_z^2 + (\alpha_x\alpha_y + \alpha_y\alpha_x) p_x p_y \\
&\quad + (\alpha_y\alpha_z + \alpha_z\alpha_y) p_y p_z + (\alpha_z\alpha_x + \alpha_x\alpha_z) p_z p_x. \qquad (20.39)
\end{aligned}$$

Substituting Eqs. (20.38) and (20.39) into Eq. (20.37), we get

$$\{E^2 - c^2 \left[\alpha_x^2 p_x^2 + \alpha_y^2 p_y^2 + \alpha_z^2 p_z^2 + (\alpha_x\alpha_y + \alpha_y\alpha_x)\, p_x p_y \right.$$
$$+ (\alpha_y\alpha_z + \alpha_z\alpha_y)\, p_y p_z + (\alpha_z\alpha_x + \alpha_x\alpha_z)\, p_z p_x \left.\right]$$
$$-mc^3 \left[(\alpha_x\beta + \beta\alpha_x)\, p_x + (\alpha_y\beta + \beta\alpha_y)\, p_y + (\alpha_z\beta + \beta\alpha_z)\, p_z\right]$$
$$-\beta^2 m^2 c^4\} \Psi = 0. \tag{20.40}$$

We require this equation to be equal to

$$\left(E^2 - c^2 p^2 - m^2 c^4\right) \Psi = 0. \tag{20.41}$$

Comparing terms in Eqs. (20.40) and (20.41), we see the following.
Since

$$p^2 = p_x^2 + p_y^2 + p_z^2, \tag{20.42}$$

we see that the second term in Eq. (20.40), which is multiplied by c^2, will be equal to p^2 if

$$\alpha_x^2 = \alpha_y^2 = \alpha_z^2 = 1, \tag{20.43}$$

and

$$\left(\alpha_x\alpha_y + \alpha_y\alpha_x\right) = 0,$$
$$\left(\alpha_y\alpha_z + \alpha_z\alpha_y\right) = 0,$$
$$\left(\alpha_z\alpha_x + \alpha_x\alpha_z\right) = 0. \tag{20.44}$$

The third term in Eq. (20.40), which is multiplied by mc^3, is absent in Eq. (20.41). Therefore,

$$\alpha_x\beta + \beta\alpha_x = 0,$$
$$\alpha_y\beta + \beta\alpha_y = 0,$$
$$\alpha_z\beta + \beta\alpha_z = 0. \tag{20.45}$$

Finally, comparing the forth term in Eq. (20.40) with Eq. (20.41), we see that

$$\beta^2 = 1. \tag{20.46}$$

Operators $\vec{\alpha}$ and β satisfying the conditions (20.43)–(20.46) are called the Dirac α matrices. Notice that the conditions (20.43)–(20.46) are the same as for the Pauli spin $\frac{1}{2}$ matrices. The operator β also satisfies the anti-commutation relations, Eq. (20.45) and $\beta^2 = 1$, so it also represents a spin.

The equation

$$\left(E - c\vec{\alpha} \cdot \vec{p} - \beta mc^2\right)\Psi = 0, \tag{20.47}$$

together with the conditions (20.43)–(20.46) is called *the Dirac equation.*

The advantage of the Dirac equation over the Schrödinger and Klein–Gordon equations is that it naturally includes the spin. In the Schrödinger and Klein–Gordon equations, the spin is not present and has to be added to the wave function manually whenever the wave function determines a system with the spin. The Dirac equation involves a system composed of N particles with spin and, as such mathematically, is in the form of $2N$ differential equations.

20.4 Dirac Spin Matrices

Dimensions of the operators $\vec{\alpha}$ and β depend on the dimension of the wave function Ψ, i.e., the number of particles (spins) involved in a given system. Since spin has two values, the operators can be represented by $2N \times 2N$ matrices.

Let us illustrate this on examples of $N = 1$ and $N = 2$ systems.

For a system composed of a single particle ($N = 1$) with spin up, $|2\rangle$, and spin down, $|1\rangle$, the components of $\vec{\alpha}$ represented in the basis of the states $|2\rangle$ and $|1\rangle$ are the Pauli spin matrices

$$\alpha_x = \begin{pmatrix} 0 & 1 \\ 1 & 0 \end{pmatrix}, \quad \alpha_y = \begin{pmatrix} 0 & -i \\ i & 0 \end{pmatrix}, \quad \alpha_z = \begin{pmatrix} 1 & 0 \\ 0 & -1 \end{pmatrix}. \tag{20.48}$$

What is the matrix form of the operator β?

Since β anti-commutes with all components of $\vec{\alpha}$ and $\beta^2 = 1$, it must be a diagonal matrix but not the unit matrix. (The unit matrix commutes, not anti-commutes, with all the matrices of $\vec{\alpha}$.) Therefore, the form of the matrix β, which satisfies that property, is

$$\beta = \begin{pmatrix} 1 & 0 \\ 0 & -1 \end{pmatrix}. \tag{20.49}$$

Consider now the case involving two particles, $N = 2$. An extension of the matrix (20.49) to the case of $N = 2$ particles is a

4×4 matrix

$$\beta = \begin{pmatrix} 1 & 0 & 0 & 0 \\ 0 & 1 & 0 & 0 \\ 0 & 0 & -1 & 0 \\ 0 & 0 & 0 & -1 \end{pmatrix}. \tag{20.50}$$

In order to have the α matrices to anti-commute with β, we write the α matrices in the following basis,
rows $\{|2\rangle_1, |1\rangle_1, |2\rangle_2, |1\rangle_2\}$ and columns $\{|2\rangle_2, |1\rangle_2, |2\rangle_1, |1\rangle_1\}$,
and find the following forms

$$\alpha_x = \begin{pmatrix} 0 & 0 & 0 & 1 \\ 0 & 0 & 1 & 0 \\ 0 & 1 & 0 & 0 \\ 1 & 0 & 0 & 0 \end{pmatrix}, \quad \alpha_y = \begin{pmatrix} 0 & 0 & 0 & -i \\ 0 & 0 & i & 0 \\ 0 & -i & 0 & 0 \\ i & 0 & 0 & 0 \end{pmatrix}, \quad \alpha_z = \begin{pmatrix} 0 & 0 & 1 & 0 \\ 0 & 0 & 0 & -1 \\ 1 & 0 & 0 & 0 \\ 0 & -1 & 0 & 0 \end{pmatrix},$$

$$\tag{20.51}$$

or shortly

$$\vec{\alpha} = \begin{pmatrix} 0 & \vec{\sigma} \\ \vec{\sigma} & 0 \end{pmatrix}. \tag{20.52}$$

20.5 Verification of the Continuity Equation

Consider now if the Dirac equation satisfies the continuity equation and if we may interpret the wave function Ψ as a probability wave function.

Consider the Dirac equation in the form

$$\frac{\hbar}{i} \frac{\partial}{\partial t} \Psi + \frac{\hbar c}{i} \vec{\alpha} \cdot \nabla \Psi + \beta m c^2 \Psi = 0. \tag{20.53}$$

Take the Hermitian conjugate

$$-\frac{\hbar}{i} \frac{\partial}{\partial t} \Psi^* - \frac{\hbar c}{i} \vec{\alpha} \cdot \nabla \Psi^* + \beta m c^2 \Psi^* = 0. \tag{20.54}$$

Multiply Eq. (20.53) from the left with Ψ^*, and Eq. (20.54) with Ψ, and next subtract both equations from each other. We then get

$$-i\hbar \left(\Psi^* \frac{\partial \Psi}{\partial t} + \Psi \frac{\partial \Psi^*}{\partial t} \right) - ic\hbar \left[\Psi^* \vec{\alpha} \cdot \nabla \Psi + \nabla \Psi^* \cdot \vec{\alpha} \Psi \right] = 0. \tag{20.55}$$

Since

$$\Psi^* \frac{\partial \Psi}{\partial t} + \Psi \frac{\partial \Psi^*}{\partial t} = \frac{\partial}{\partial t} |\Psi|^2, \tag{20.56}$$

and

$$\Psi^* \vec{\alpha} \cdot \nabla \Psi + \nabla \Psi^* \cdot \vec{\alpha} \Psi = \nabla \cdot (c \Psi^* \vec{\alpha} \Psi), \tag{20.57}$$

we can write Eq. (20.55) as

$$\frac{\partial}{\partial t} |\Psi|^2 + \nabla \cdot (c \Psi^* \vec{\alpha} \Psi) = 0, \tag{20.58}$$

which is in the form of the continuity equation. Hence,

$$|\Psi|^2 = \rho \quad \text{and} \quad c \Psi^* \vec{\alpha} \Psi = \vec{J}. \tag{20.59}$$

Notice that both ρ and $\vec{J}$ are real and $\rho \geq 0$. Thus, we can interpret ρ as the probability density. Therefore, we may conclude that the Dirac equation is a relativistic form of the Schrödinger equation for the probability wave function.

Tutorial Problems

Problem 20.1 Show that the Klein–Gordon equation for a free particle is invariant under the Lorentz transformation. The Lorentz transformation is given by

$$x' = \gamma (x - \beta ct),$$
$$y' = y,$$
$$z' = z,$$
$$ct' = \gamma (ct - \beta x),$$

where $\gamma = (1 - \beta^2)^{-1/2}$ is the Lorentz factor, $\beta = u/c$, and u is the velocity with which an observer moves.

Problem 20.2 Act on the Dirac equation

$$(E - c\vec{\alpha} \cdot \vec{p} - \beta mc^2) \Psi = 0$$

with the operator

$$E + c\vec{\alpha} \cdot \vec{p} + \beta mc^2$$

to find under which conditions the Dirac equation satisfies the relativistic energy relation

$$E^2 = c^2 p^2 + m^2 c^4.$$

Here, $\vec{\alpha} = \alpha_x \hat{i} + \alpha_y \hat{j} + \alpha_z \hat{k}$ is a three-dimensional Hermitian operator and β is a one-dimensional Hermitian operator. The operator β does not commute with any of the components of $\vec{\alpha}$.

Problem 20.3 Verify the anti-commutation relations (20.45) for a system composed of two particles $N = 2$, for which β is given by Eq. (20.50) and α matrices are given in Eq. (20.51).

Chapter 21

Systems of Identical Particles

In this final chapter, we study the properties of systems of identical particles. We will find wave functions of a system composed of N identical and independent particles. From the properties of the wave functions, we will find that identical particles can be characterized by symmetric as well as antisymmetric wave functions. As a result, we will distinguish two kinds of particles: bosons and fermions.

Consider a system composed of N parts (subsystems), e.g., a system of N identical and independent particles, whose Hamiltonian is given by

$$\hat{H} = \sum_{i=1}^{N} \hat{H}_i \, , \tag{21.1}$$

and the wave function is

$$\Psi(r) = \phi_1(r_1)\phi_2(r_2)\ldots\phi_N(r_N) \, , \tag{21.2}$$

where $\phi_i(r_i)$ is the wave function of the ith particle located at the point r_j, or equivalently we can say that $\phi_i(r_j)$ is the wave function of the jth particle being in the ith state.

However, the wave function $\Psi(r)$ is not the only eigenfunction of the system. For example, a wave function

$$\Psi(r) = \phi_1(r_2)\phi_2(r_1)\ldots\phi_N(r_N) \, , \tag{21.3}$$

Quantum Physics for Beginners
Zbigniew Ficek
Copyright © 2016 Pan Stanford Publishing Pte. Ltd.
ISBN 978-981-4669-38-2 (Hardcover), 978-981-4669-39-9 (eBook)
www.panstanford.com

is also an eigenfunction of the system with the same eigenvalue.

Proof: Consider the eigenvalue equation with the eigenfunction (21.2):

$$\hat{H}\,\Psi(r) = \sum_{i=1}^{N} \hat{H}_i \Psi(r)$$

$$= \sum_{i}^{N} \hat{H}_i \phi_1(r_1)\phi_2(r_2)\dots\phi_N(r_N) = \sum_{i=1}^{N} E_i \Psi(r)\,.$$

Consider now the eigenvalue equation with the eigenfunction (21.3):

$$\hat{H}\,\Psi(r) = \sum_{i=1}^{N} \hat{H}_i \Psi(r) = \sum_{i}^{N} \hat{H}_i \phi_1(r_2)\phi_2(r_1)\dots\phi_N(r_N)\,.$$

Since

$$\hat{H}_1\phi_1(r_2) = E_1\phi_1(r_2)\,, \quad \text{and} \quad \hat{H}_2\phi_2(r_1) = E_2\phi_2(r_1)\,,$$

we get $\hat{H}\,\Psi(r) = \sum_{i=1}^{N} E_i \Psi(r)$.

Even if

$$\hat{H}_1\phi_1(r_2) = E_2\phi_1(r_2)\,, \quad \text{and} \quad \hat{H}_2\phi_2(r_1) = E_1\phi_2(r_1)\,,$$

we get

$$\hat{H}\,\Psi(r) = \sum_{i=1}^{N} E_i \Psi(r)\,, \tag{21.4}$$

as required.

In fact there are $N!$ permutations of $\phi_i(r_j)$, which are eigenfunctions of the system. Moreover, an arbitrary linear combination of the wave functions $\phi_i(r_j)$ is also an eigenfunction of the system. We will illustrate this for $N = 2$.

21.1 Symmetric and Antisymmetric States

Consider an arbitrary linear combination of two wave functions

$$\Psi(r) = \frac{1}{\sqrt{|a|^2 + |b|^2}}\,[a\Psi(r_{12}) + b\Psi(r_{21})]\,, \tag{21.5}$$

where $\Psi(r_{12} = \phi_1(r_1)\phi_2(r_2)$ and $\Psi(r_{21} = \phi_1(r_2)\phi_2(r_1)$. Then

$$\hat{H}\Psi(r) = \left(\hat{H}_1 + \hat{H}_2\right)\Psi(r)$$

$$= \frac{1}{\sqrt{|a|^2 + |b|^2}}\left[a\,(E_1 + E_2)\,\Psi(r_{12}) + b\,(E_1 + E_2)\,\Psi(r_{21})\right]$$

$$= (E_1 + E_2)\Psi(r). \tag{21.6}$$

We know that in the linear combination, $|a|^2/(|a|^2 + |b|^2)$ is the probability that the particle "1" is at the position r_1 and the particle "2" is at r_2. Equivalently, for $r_1 = r_2$, we can say that this is the probability that the particle "1" is in a state $|1\rangle$, and the particle "2" is in a state $|2\rangle$.

Similarly, $|b|^2/(|a|^2 + |b|^2)$ is the probability that the particle "1" is at the position r_2, and the particle "2" is at r_1.

Note that in general, the probabilities are different. However, for two identical particles, the probabilities should be the same as we cannot distinguish between two identical particles.

Thus, for two identical particles, $|a| = |b|$. Hence, the parameters a and b can differ only by a phase factor: $b = a\,\exp(i\phi)$, where ϕ is a real number:

$$\Psi(r) = \frac{1}{\sqrt{2}}\left[\Psi(r_{12}) + e^{i\phi}\Psi(r_{21})\right]. \tag{21.7}$$

If we exchange the positions of the particles ($r_1 \leftrightarrow r_2$) or energy states ($|1\rangle \leftrightarrow |2\rangle$), then we obtain

$$\Psi'(r) = \frac{1}{\sqrt{2}}\left[e^{i\phi}\Psi(r_{12}) + \Psi(r_{21})\right]. \tag{21.8}$$

Thus, the exchange of $r_1 \leftrightarrow r_2$ or $|1\rangle \leftrightarrow |2\rangle$ is equivalent to multiplying $\Psi(r)$ by $e^{i\phi}$ and taking $e^{2i\phi} = 1$. Hence

$$e^{i\phi} = \pm 1, \tag{21.9}$$

and therefore the wave functions of identical particles are either symmetrical or antisymmetrical

$$\Psi_s(r) = \frac{1}{\sqrt{2}}\left[\Psi(r_{12}) + \Psi(r_{21})\right],$$

$$\Psi_a(r) = \frac{1}{\sqrt{2}}\left[\Psi(r_{12}) - \Psi(r_{21})\right]. \tag{21.10}$$

Note that

$$\Psi_s(r_{12}) = \Psi_s(r_{21}),$$

$$\Psi_a(r_{12}) = -\Psi_a(r_{21}). \tag{21.11}$$

Properties of symmetrical and antisymmetrical functions

Property 1. If $\hat{H} = \hat{H}_{12} = \hat{H}_{21}$, i.e., the Hamiltonian is symmetrical, then $\hat{H}\Psi(r)$ has the same symmetry as $\Psi(r)$.

Proof: Take $\Psi(r) = \Psi_s(r)$. Denote $f_{12} = \hat{H}_{12}\Psi_s(r_{12})$, then

$$f_{21} = \hat{H}_{21}\Psi_s(r_{21}) = \hat{H}_{12}\Psi_s(r_{12}) = f_{12}.$$

Take now $\Psi(r) = \Psi_a(r)$. Then

$$f_{21} = \hat{H}_{21}\Psi_a(r_{21}) = \hat{H}_{12}\left(-\Psi_s(r_{12})\right) = -f_{12},$$

as required.

Property 2. Symmetry of the wave function does not change in time, i.e., the wave function initially symmetrical (antisymmetrical) remains symmetrical (antisymmetrical) for all times.

Proof: Consider the evolution of a wave function $\Psi(t)$ in a time dt:

$$\Psi(t + dt) = \Psi(t) + \frac{\partial\Psi}{\partial t}dt.$$

Thus, symmetry of the wave function depends on the symmetry of $\partial\Psi/\partial t$. From the time-dependent Schrödinger equation

$$i\hbar\frac{\partial\Psi}{\partial t} = \hat{H}\Psi,$$

we see that $\partial\Psi/\partial t$ has the same symmetry as $\hat{H}\Psi$. From the property **1**, we know that $\hat{H}\Psi$ has the same symmetry as Ψ. Therefore, $\Psi(t + dt)$ has the same symmetry as $\Psi(t)$, as required.

Difference between symmetric and antisymmetric functions

Antisymmetric function can be written in a form of a determinant, called the *Slater determinant*:

$$\Psi_a(r) = \frac{1}{\sqrt{N!}}\begin{vmatrix} \phi_1(r_1) & \phi_1(r_2) & \cdots & \phi_1(r_N) \\ \phi_2(r_1) & \phi_2(r_2) & \cdots & \phi_2(r_N) \\ & \cdot & & \\ & \cdot & & \\ & \cdot & & \\ \phi_N(r_1) & \phi_N(r_2) & \cdots & \phi_N(r_N) \end{vmatrix}, \qquad (21.12)$$

where $1/\sqrt{N!}$ is the normalization constant.

If two particles are at the same point, $r_1 = r_2$, and then two columns of the determinant (21.12) are equal, giving $\Psi_a(r) = 0$. Thus, two particles determined by the antisymmetric function cannot be at the same point. Similarly, if two particles are in the same state, $\phi_1(r_1) = \phi_1(r_2)$, and again two columns are equal giving $\Psi_a(r) = 0$.

A symmetrical function cannot be written in the form of a determinant. Thus, particles that are determined by symmetrical functions can be in the same point or in the same state.

Hence, particles can be divided into two types: those determined by antisymmetric functions (called *fermions*) and those determined by symmetrical functions (called *bosons*).

Examples of fermions and bosons

Fermions: electrons, protons, neutrons, neutrinos.
Bosons: photons, phonons, π mesons, α particles.

From experiments, we know that we can distinguish between fermions and bosons by looking at their spins. Fermions have half-integer spin ($1/2, 3/2$, etc., in units of $\hbar$), whereas bosons have integer spin ($0, 1, 2$, etc.).

Since an arbitrary number of bosons can occupy the same state, atoms with integer spin (atomic bosons) can abruptly condensate into a single ground state when the temperature of the atoms goes below a certain critical value. We call this process *Bose–Einstein condensation.*[a]

21.2 Pauli Principle

In atoms, a limited number of electrons (fermions) can occupy the same energy level. How many electrons does it take to fill an energy

---

[a]Readers wishing to learn more about Bose–Einstein condensation are referred to a book by L. Pitaevskii and S. Stringari, *Bose–Einstein Condensation* (Clarendon Press, Oxford, 2003).

level? The answer to this question is given by the Pauli principle,[a] also called exclusion principle.

Pauli principle

No two electrons can have the same quantum numbers (n, l, m, s) in a multi-electron atom.

It is also known as the exclusion principle, for the simple reason that if an electron has the quantum numbers $(nlms)$, then at least one of the quantum numbers of any further electrons must be different.

In an atom, for a given n, there are $2(2l + 1)$ degenerate states corresponding to $l = 0, 1, 2, \ldots, m = -l, \ldots, l$, and $s = -\frac{1}{2}, +\frac{1}{2}$. Thus, for a given n, the total number of electrons in the energy state Ψ_{nlm} is

$$\sum_{l=-m}^{l=m} 2(2l + 1) = 2n^2 . \qquad (21.13)$$

Following the Pauli principle, we can find the numbers of electrons in the energy states

$1s$,	$2s$,	$2p$,	$3s$,	$3p$,	$4s$,	$3d$,	$4p$,	$5s$,	$4d$,	$5p$
2	2	6	2	6	2	10	6	2	10	6

The Pauli principle prevents the energy states being occupied by an arbitrary number of electrons. The state $1s$ can be occupied by two electrons. Hence, as more electrons are added, the energy of the atom grows along with its size. Thus, the Pauli principle prevents all atoms having the same size and the same energy. This is the quantum physics explanation of atomic sizes and energies.

Since the number of electrons on given energy levels is limited, we get different ground state configurations for different atoms. The ground state of a many-electron atom is that in which the electrons occupy the lowest energy levels that they can occupy.

If the number of electrons for a given nl is $2(2l + 1)$, we say that there is a *closed shell*. Examples: Helium, Beryllium, Neon.

---

[a] Pauli was granted the Nobel Prize in 1945 for the discovery of the exclusion principle (also called Pauli principle).

Since the chemical properties of atoms depend on the number of electrons outside the closed shells, the atoms with similar outer configurations will have similar chemical properties. Examples: The Alkali metals: lithium $(1s)^2 2s$, sodium $(1s)^2 (2s)^2 (2p)^6 3s$, and potassium $(1s)^2 (2s)^2 (2p)^6 (3s)^2 (3p)^6 4s$.

This is the explanation from quantum physics of the periodic structure of the elements.

21.3 Symmetric and Antisymmetric States of $N > 2$ Particles

For a system composed of two particles, the question of which of the two possible functions Ψ_{12} and Ψ_{21} should contribute to the antisymmetric combination with the positive and which with the negative sign is not a problem. However, for $N > 2$ particles, it is a problem which wave functions contribute with the positive and which with the negative sign. There is a rule for choosing the sign, which we will illustrate on an example of $N = 3$ particles.

Consider a system of $N = 3$ independent particles. The number of possible permutations is $3! = 6$, so that we have 6 possible wave functions

$$\Psi_{123}, \ \Psi_{231}, \ \Psi_{312}, \ \Psi_{213}, \ \Psi_{132}, \ \Psi_{321}. \tag{21.14}$$

The symmetric function of the three particles

$$\Psi_s = \frac{1}{\sqrt{3!}} \left(\Psi_{123} + \Psi_{231} + \Psi_{312} + \Psi_{213} + \Psi_{132} + \Psi_{321} \right). \tag{21.15}$$

The antisymmetric function of the particles

$$\Psi_a = \frac{1}{\sqrt{3!}} \left[(\Psi_{123} + \Psi_{231} + \Psi_{312}) - (\Psi_{213} + \Psi_{132} + \Psi_{321}) \right]. \tag{21.16}$$

For the symmetric function, all six terms have the positive sign so that they can be added in an arbitrary order. For the antisymmetric function, three terms have the positive sign and three terms have the negative sign. A question then arises: Which functions should be contributing with the positive and which with the negative sign? The rule of finding the sign of a particular term is as follows.

The rule is how many times do we have to change a pair of numbers until the sequence 123 is obtained. For example:

$$(231) \rightarrow (213) \rightarrow (123) \qquad (21.17)$$

so that two changes were made to get the sequence (123). Since the number of changes was an even number, the sign of the term is chosen $(+)$. If the number of changes is an odd number, the sign of the term is chosen $(-)$. For example, take the term (321):

$$(321) \rightarrow (312) \rightarrow (132) \rightarrow (123) \qquad (21.18)$$

so we had to make 3 changes of the numbers, an odd number of changes. The sign of the term is, therefore, $(-)$.

21.4 Identical (Nondistinguishable) and Nonidentical (Distinguishable) Particles: Degeneracy of Wave Function

Consider a system of N independent particles each of energy E_i determined by the integer number n_i.

For example, if a particle is represented by harmonic oscillators, then $E_i = n_i \hbar \omega = n_i E_0$. Total energy is $E = (n_1 + n_2 + \ldots) E_0$, and the wave function of the system is

$$\Psi_{n_1, n_2, n_3, \ldots} = \phi_1(n_1) \phi_2(n_2) \phi_3(n_3) \cdots \qquad (21.19)$$

We see that the energy E and the wave function Ψ are specified by a set of integer numbers $(n_1, n_2, n_3, \ldots)$. Thus, there might be few wave functions corresponding to the same energy E. If it happens, we say that the energy level is *degenerate*, and the number of states corresponding to this energy is called *degeneracy*.

Worked Example

Consider a system of three identical and independent harmonic oscillators. If the oscillators are in their lowest energy states, for which $n_1 = n_2 = n_3 = 1$, the total energy of the oscillators is $E = 3E_0$ and there is only one combination of the numbers $(n_1, n_2, n_3) =$

(1, 1, 1). Thus, there is only one wave function corresponding to this energy, $\Psi_{1,1,1}$. We say that the level has degeneracy 1.

If the oscillators have the total energy $E = 4E_0$, there are three combinations of n_1, n_2, and n_3 which sum to 4. These combinations are $(n_1 = 2, n_2 = 1, n_3 = 1)$, $(n_1 = 1, n_2 = 2, n_3 = 1)$, and $(n_1 = 1, n_2 = 1, n_3 = 2)$. Thus, there are three wave functions corresponding to this energy: $\Psi_{2,1,1}$, $\Psi_{1,2,1}$, and $\Psi_{1,1,2}$. In this case, we say that the energy level has degeneracy 3.

Tutorial Problems

Problem 21.1 Consider a system of three identical and independent particles.

(a) What would be the level of degeneracy if the particle 1 of energy $n_1 = 2$ would be distiguished from the other two particles?
(b) What would be the level of degeneracy if the distinguished particle has energy $n_1 = 1$?

Problem 21.2 Two identical particles of mass m are in the one-dimensional infinite potential well of a dimension a. The energy of each particle is given by

$$E_i = n_i^2 \frac{\pi^2 \hbar^2}{2ma^2} = n_i E_0.$$ (21.20)

(a) What are the values of the four lowest energies of the system?
(b) What is the degeneracy of each level.

Problem 21.3 Redistribution of particles over a finite number of states

(a) Assume we have n identical particles, which can occupy g identical states. The number of possible distributions, if particles were bosons, is given by the number of possible permutations

$$t = \frac{(n + g - 1)!}{n!(g - 1)!}$$ (21.21)

For example, $n = 3$ and $g = 2$ give $t = 4$. However, this is true only for identical bosons. What would be the number of possible redistributions if the particles were fermions or were distinguishable?

(b) Find the number of the allowed redistributions if the particles were:

   (i) Identical bosons
  (ii) Identical fermions
 (iii) Nonidentical fermions
 (iv) Nonidentical bosons

Illustrate this on an example of $n = 2$ independent particles that can be redistributed over 5 different states.

Final Remark

Although this textbook focused on a small part of quantum physics, it should nevertheless be appropriate to close by emphasizing the importance of quantum phenomena in the development of new areas in science and technology. The predictions of quantum physics have turned research and technology into new directions and have led to numerous technological innovations and the development of a new technology on the scale of single atoms and electrons, called quantum technology or nanotechnology. The ability to manufacture and control the dimensions of tiny structures, such as quantum dots, allows us to engineer the unique properties of these structures and predict new devices such as quantum computers. A quantum computer can perform mathematical calculations much faster and store much more information than a classical computer by using the laws of quantum physics. The technology for creating a quantum computer is still in its infancy because it is extremely difficult to control quantum systems, but it is developing very rapidly with little sign of the progress slowing.

We have seen in our journey through the backgrounds of quantum physics that despite its long history, quantum physics still challenges our understanding and continues to excite our imagination. Feynman, in his lectures on quantum physics, referred in the following way to our understanding of quantum physics:

> *I think I can safely say that nobody understands quantum mechanics.*

In summary of the book, I think I can safely say:

> *If you think you now know quantum physics, it means you do not know anything.*

Appendix A

Derivation of the Boltzmann distribution function P_n

Assume that we have n identical particles (e.g., photons), each of energy E, which can occupy g identical states. The number of possible distributions of n particles between g states is given by

$$t = \frac{(n + g - 1)!}{n!(g - 1)!} .$$ \hfill (A.1)

For example: $n = 3$, $g = 2$ gives $t = 4$, as it is illustrated in Fig. A.1.

We will find maximum of t with the condition that $nE = $ constant, where E is the energy of each particle.

Taking ln of both sides of Eq. (A.1), we get

$$\ln t = \ln(n + g - 1)! - \ln n! - \ln(g - 1)! .$$ \hfill (A.2)

Using Sterling's formula

$$\ln n! = n \ln n - n ,$$ \hfill (A.3)

Figure A.1 Example of possible distributions of three particles between two states.

and assuming that $g \gg 1$, i.e., $g - 1 \approx g$, we obtain

$$\ln t = g \ln \frac{n+g}{g} + n \ln \frac{n+g}{n} . \tag{A.4}$$

We find maximum of $\ln t$ using the method of *Lagrange undetermined multipliers*. In this method, we construct a function

$$K = \ln t - \lambda n E , \tag{A.5}$$

where λ is called a *Lagrange undetermined multiplier*, and find the extremum

$$\frac{\partial K}{\partial n} = 0 . \tag{A.6}$$

Thus, we get

$$\ln \frac{n}{n+g} + \lambda E = 0 , \tag{A.7}$$

from which, we find

$$n = \frac{g}{e^{\lambda E} - 1} . \tag{A.8}$$

This is the Bose–Einstein distribution function. Since n is dimensionless, λ should be inverse of energy. We choose $\lambda = 1/(k_B T)$, where $k_B T$ is the energy of free, noninteracting particles. When $g/n \gg 1$, we can approximate Eq. (A.8) by

$$n = g e^{-\frac{E}{k_B T}} , \tag{A.9}$$

which is known in statistical physics as the Boltzmann distribution. This gives the number of particles n of energy E.

If there are particles, among N particles, that can have different energies E_i, then

$$\frac{n_i}{N} = \frac{g}{N} e^{-\frac{E}{k_B T}} \tag{A.10}$$

is the probability that n_i particles of the total N particles have energy E_i.

Thus, we can write

$$P_n = a e^{-\frac{E}{k_B T}} , \tag{A.11}$$

where a is a constant.

Since the probability is normalized to one ($\sum_n P_n = 1$), we finally get

$$P_n = \frac{e^{-\frac{E}{k_B T}}}{\sum_n e^{-\frac{E}{k_B T}}} . \tag{A.12}$$

The sum $\sum_n e^{-\frac{E}{k_B T}}$ is called the *partition function*.

Appendix B

Useful Mathematical Formulae

Some useful properties of trigonometrical functions:

$$\sin(\alpha \pm \beta) = \sin\alpha\cos\beta \pm \sin\beta\cos\alpha$$

$$\cos(\alpha \pm \beta) = \cos\alpha\cos\beta \mp \sin\alpha\sin\beta$$

$$\sin^2\alpha = \frac{1}{2}(1 - \cos 2\alpha)$$

$$\cos^2\alpha = \frac{1}{2}(1 + \cos 2\alpha)$$

$$\cot\frac{\alpha}{2} = \frac{\sin\alpha}{1 - \cos\alpha}$$

$$\int_0^\pi \sin^3\theta\, d\theta = \frac{4}{3}$$

$$\int_0^{2\pi} \sin(m\phi)\sin(n\phi)\, d\phi = \begin{cases} 0 & \text{for } m \neq n \\ \pi & \text{for } m = n \end{cases}$$

$$\int_0^{2\pi} \cos(m\phi)\cos(n\phi)\, d\phi = \begin{cases} 0 & \text{for } m \neq n \\ \pi & \text{for } m = n \end{cases}$$

$$\int_0^{2\pi} \sin(m\phi)\cos(n\phi)\, d\phi = 0 \quad \text{for all} \quad m \text{ and } n$$

Useful integral expressions

$$\int_0^\infty \frac{x^3}{e^x - 1} dx = \frac{\pi^4}{15}, \qquad \int_{-\infty}^\infty e^{-\alpha x^2} dx = \sqrt{\frac{\pi}{\alpha}},$$

$$\int_{-\infty}^\infty x e^{-\alpha x^2} dx = 0, \qquad \int_{-\infty}^\infty x^2 e^{-\alpha x^2} dx = \frac{1}{2\alpha} \sqrt{\frac{\pi}{\alpha}}.$$

$$\int_0^\infty r^n e^{-\alpha r} dr = \frac{n!}{\alpha^{n+1}},$$

from which, we find

$$\int_0^\infty e^{-\alpha r} dr = \frac{1}{\alpha}, \qquad \int_0^\infty r e^{-\alpha r} dr = \frac{1}{\alpha^2},$$

$$\int_0^\infty r^2 e^{-\alpha r} dr = \frac{2}{\alpha^3}, \qquad \int_0^\infty r^3 e^{-\alpha r} dr = \frac{6}{\alpha^4}$$

Taylor series

$$\omega_k = \omega_{k_0 + \beta} = \omega_0 + \left(\frac{d\omega}{d\beta}\right)_{k_0} \beta + \frac{1}{2} \left(\frac{d^2\omega}{d\beta^2}\right)_{k_0} \beta^2 + \dots$$

$$e^{\pm x} = 1 \pm x + \frac{x^2}{2!} \pm \frac{x^3}{3!} + \dots$$

$$\sin x = x - \frac{x^3}{3!} + \frac{x^5}{5!} - \frac{x^7}{7!} + \dots$$

$$\cos x = 1 - \frac{x^2}{2!} + \frac{x^4}{4!} - \frac{x^6}{6!} + \dots$$

Kronecker δ function

$$\delta_{mn} = \begin{cases} 1 & \text{if } m = n \\ 0 & \text{if } m \neq n. \end{cases}$$

Dirac delta function

$$\delta(x) = \begin{cases} 0 & \text{if } x \neq 0 \\ \infty & \text{if } x = 0, \end{cases} \qquad \text{such that} \quad \int_{-\infty}^\infty f(x)\delta(x)dx = f(0),$$

for any function $f(x)$.

Appendix C

Physical Constants and Conversion Factors

Bohr magneton	$m_B = 9.724 \times 10^{-24}$ [J/T]
Bohr radius	$a_o = 5.292 \times 10^{-11}$ [m]
Boltzmann constant	$k_B = 1.381 \times 10^{-23}$ [J/K]
charge of an electron	$e = -1.602 \times 10^{-19}$ [C]
permeability of vacuum	$\mu_0 = 4\pi \times 10^{-7}$ [H/m]
permittivity of vacuum	$\varepsilon_0 = 8.854 \times 10^{-12}$ [F/m]
Planck's constant	$h = 6.626 \times 10^{-34}$ [J.s]
	$= 4.14 \times 10^{-15}$ [eV.s]
(Planck's constant)$/2\pi$	$\hbar = 1.055 \times 10^{-34}$ [J.s]
	$= 6.582 \times 10^{-16}$ [eV.s]
rest mass of electron	$m_e = 9.110 \times 10^{-31}$ [kg]
rest mass of proton	$m_p = 1.673 \times 10^{-27}$ [kg]
Rydberg constant	$R = 1.097 \times 10^7$ [m^{-1}]
speed of light in vacuum	$c = 2.9979 \times 10^8$ [m/s]
Stefan–Boltzmann constant	$\sigma \doteq 5.670 \times 10^{-8}$ [W/m$^2 \cdot$ K^4]

$1\,\text{Å} = 10^{-10}$ [m] ; $1\,\text{fm} = 10^{-15}$ [m] ; $1\,\text{eV} = 1.602 \times 10^{-19}$ [J]

$1\,\text{J} = 6.241 \times 10^{18}$ [eV] ; $\pi = 3.142$; $e = 2.718$.

Index